Context Changes Everything

Context Changes Everything

How Constraints Create Coherence

Alicia Juarrero

The MIT Press
Cambridge, Massachusetts
London, England

This book was set in Sabon by Westchester Publishing Services. Printed and bound in the United States of America.

Library of Congress Cataloging-in-Publication Data

Names: Juarrero, Alicia, author.
Title: Context changes everything : how constraints create coherence / Alicia Juarrero.
Description: Cambridge, Massachusetts : The MIT Press, [2023] | Includes bibliographical references and index.
Identifiers: LCCN 2022030581 (print) | LCCN 2022030582 (ebook) | ISBN 9780262545662 | ISBN 9780262374781 (epub) | ISBN 9780262374774 (pdf)
Subjects: LCSH: Complexity (Philosophy) | Causation.
Classification: LCC B105.C473 J83 2023 (print) | LCC B105.C473 (ebook) | DDC 117—dc23/eng/20230124
LC record available at https://lccn.loc.gov/2022030581
LC ebook record available at https://lccn.loc.gov/2022030582

10 9 8 7 6 5 4

This book is dedicated to Julia, Marama, Ana, Alejandro, Miguel, Kiri, Rewa, Mahina, and José Raúl, purveyors of context extraordinaire.

Contents

I

1

What Went Wrong? The Backstory

The controversy surrounding the concept of identity illustrates how philosophical and scientific presuppositions that go without saying entangle us in never-ending mazes.

In the history of ideas, claims that *identity* consists of a defining essence have a long backstory. In fact, one could arguably claim that it is among the oldest backstories in philosophy, a backstory that involves the paradox of the One and the Many. What makes different entities (e.g., dachshunds, boxers, chihuahuas, Great Danes) the same, that is, dogs? What makes *anything* continue being the same thing? How is it that something can persist as itself—despite changes wrought by either developmental processes, the environment, or internal malfunction?

Issues with identity are not new; the problem was first posed in the sixth century BCE. The classical version went as follows: If all the planks of the ship of Theseus are replaced over time, is it still "the ship of Theseus," even when not one of the original planks remains? Does anything persist as the same thing, the ship of Theseus, despite such significant alterations? How is change compatible with remaining the same? What does *same* mean, in this context or any other?

In 2019, Chinese scientists inserted a human gene into the brains of eleven monkeys; those monkeys then outperformed normal monkeys in tests of short-term memory. Even more striking, the modified monkeys' brain developed into an organ more like a human's than like that of the unmodified monkeys (Regalado 2019). Are these animals now partly human? In 2021, an international team of scientists injected human stem cells into primates (Subbaraman 2021). The interspecies chimeras grew for two weeks before being destroyed. If a human being's body parts are gradually replaced with silicon components until no original organic material remains, would they still be human, or a cyborg? If all my memories disappeared, or if my beliefs suddenly became entirely different, would I still be me? What is it to be me? Just as the Ship of Theseus is one of the oldest backstories in philosophy, tales of aliens taking over the bodies of earthlings are among the oldest themes in science fiction.[1]

Today's controversy about identity is not just an old story; it is also an oft-told story (Toulmin 1990; Juarrero 1999). However, the central theme of this book is not identity, nor will this book be solely a survey of these debates. The problem of identity happens to be a noteworthy philosophical topic that encapsulates in a single controversy three categories that have been egregiously responsible for framing our views of reality in a misleading manner. Our understanding of many concepts, not only those that arise with respect to identity, continues to be misdirected by outdated and erroneous ideas about three central issues:

- Are wholes different from aggregates?
- Are context and history part of the fabric of reality?
- How do cause and effect work?

The three questions are so deeply interwoven in the way we see things that a discussion of one inevitably brings to bear the others. The problem of wholes and aggregates is the problem of Types or Kinds and Essences, the problem of context and history is the problem of Interactions and Relations, and the problem of causes and effects is, well, the huge philosophical problem of Causation.

The last one, Causation, is the most recalcitrant and pervasive of the three. It wraps around the other two: Do interactions among individual entities create real novelty in the form of coherent wholes? Can the context in which an event occurs have any influence either on whether it happens at all or on exactly what happens?

This chapter presents highlights from the history of these ideas. It describes how relations and interactions, context and history, came to be thought of as causally impotent; how seemingly coherent totalities were reduced to nothing but aggregates; and how the idea of cause and effect came to be restricted to energy-transferring processes. These controversies are problematic, not only because of what they have to say about the nature of identity; they mainly arise because of our deeply flawed views about *causality* and *coherence*—about the causes of coherence. Once we understand what went wrong with these two notions, we can begin to reimagine a different and more inclusive interpretive framework that can also make room for identity as contextually embedded coherence and interdependence. The book takes preliminary steps in that direction.

Types, Kinds, and Essences

The fable of the Scorpion and the Frog shows how "identity" has traditionally been thought to capture the essence that uniquely distinguishes something from everything else—everything else that something is not. The scorpion asks the frog to ferry it across the river safely. When the frog hesitates, fearful the scorpion would kill it, the scorpion protests that doing so would mean that both would drown. The frog accepts that reasoning. Halfway across the river the scorpion stings the frog. As they are both drowning, the dying frog asks the scorpion, "Why did you do that?" To which the scorpion replies, "I couldn't help it. It's my nature." Thinking of identity in this way suggests that if certain properties, internal to the creature, were missing, it would no longer be itself because those properties define what it is to be a scorpion. They capture, that is, a thing's fundamental essence and therefore its identity.

The idea of nature or essence as internal and standalone is noteworthy because of the unacknowledged assumptions it uncritically accepts. Essence as internal represents a refusal to acknowledge that interactions and relations play a role in a thing's nature; it also refuses to recognize that relational properties like coordination, integration, and context embeddedness are real. It ignores both the past and current circumstances. It underpins, in short, a worldview that dismisses time and place—context in general—from reality. These become passive containers instead. Such failures make it impossible to understand coherence and identity.

That backstory goes as follows.

Beginning with Thales in the sixth century BCE, pre-Socratic philosophers embarked on a search for the essence and nature of reality. What made pre-Socratics different from their precursors was their faith in reason and logic, as opposed to an unquestioning acceptance of mythic tales (Juarrero 1993). Is Water the foundation of reality, asked Thales? Is there more than one fundamental essence—perhaps, air, wind, fire, as well as water? Maybe all four, as Empedocles proposed? Or is the essence of reality change (Heraclitus)? Or an unchanging Plenum (Parmenides)? Lucretius, the first recorded *atomist*, noted that whatever turns out to be fundamental must be in-divisible—*a-tom*. Otherwise, it could be decomposed into more simple stuff and so would not be fundamental. The late Carl Sagan, host of the popular public television series *Cosmos*, lamented that the history of ideas went wrong when Socrates veered from this pre-Socratic interest in cosmological and natural questions and turned instead to ethical and normative ones.

Fast forward to the eighteenth century. Modern English atomist John Dalton noticed that chemicals could be combined into more complex substances or broken down into more fundamental elements. In a nod toward Lucretius, Dalton called the fundamental particles *atoms* and identified weight as their primary property. In that same century, French chemist Antoine Lavoisier added mass as a second primary property of atoms.

Primary and Secondary (Accidental) Properties

Modern science and philosophy consolidated the idea of primary properties as the ground of reality and therefore of identity; only primary properties constitute the essence of things. One of the features that allegedly makes primary properties primary is their ability to exist on their own, independent of others; they do not depend on anything else either to exist or to be perceived. Properties like atomic weight and mass—those that underlie solidity, extension, and quantity—are often listed as preeminent examples of primary properties. From this perspective, essence and identity are what might be called *internalist* concepts: they pick out properties that are internal to the entity in question; they are also objectively measurable and observer independent.

Once the internalist definition was adopted, concluding that interactions contribute nothing fundamental to reality followed. Relations with other entities in the environment and with the past bring context to bear, but because context was viewed as a container external to the thing's essence, it too was set aside as irrelevant to identity. The philosophical conundrum posed by the measurement problem in quantum physics originates in this framing.

Contrast the seeming ontological independence and self-sufficiency of atomic mass and weight with the context dependence of color and sound. It has been known since classical times (Democritus) that flavors, odors, color, sounds, and even temperature are inherently relational and contextual. Inserting a hand in a bucket of tepid water feels cold if it comes after inserting the hand in hot water, but it feels warm if it comes after icy water. Because they are not independent of perception, relational properties dependent on an observer were considered subjective.[2] In the language of academic philosophy, properties that arise from interactions, especially with human observers, are secondary or accidental properties. For that very reason, they are also ontologically secondary, their reality derivative.

In natural philosophy, *Kind* traditionally refers to natural kinds and is often used synonymously with *Type* and *Universals* to mean classes of entities or processes defined by common primary properties.[3] "It's my [scorpion] nature" follows that reasoning. On this view, kinds and types pick out real, universal features in the world. *Scorpionality*, if you will, is real, the set of primary properties that makes scorpions *scorpions*. The elements of the periodic table are prototypes of natural kinds not least because they represent measurable and observer-independent properties such as the number of protons in the nucleus. All instances of element X have Y number of protons in their nucleus. This primary property constitutes their essence.

It is important to note that, even in classical times, types, kinds, and universals were understood to be multiply realized: differently embodied tokens can realize the same type. Some actual scorpions are ground dwelling; others are tree dwellers, but as scorpions, they possess the same basic primary properties. Likewise, whether isosceles, equilateral, or scalene, actual triangles are *tokens* of the type triangularity. They all realize its essential, primary properties: three-sided, enclosed, two-dimensional figures whose interior angles sum to 180 degrees. They would not be triangles if they did not.

According to Plato, even if material tokens of types ceased to exist, if this triangle I drew in the sand got washed away with the tide, the essence of triangularity would continue to exist even if no longer instantiated. Platonic Forms such as Justice, Goodness, Truth, and Beauty as well as triangularity and numbers were thought to exist in a transcendental realm, fixed and eternal. They were considered real even if no actual triangles, or written numbers, or actual cases of justice, goodness, truth, or beauty had ever existed.

So, Carl Sagan did not get it quite right: Socrates was not entirely deviating from the tradition of the pre-Socratics. Platonic dialogues such as the *Republic* (about justice) and *Meno* (about knowledge acquisition) narrate Socrates' attempts to elucidate and define the universal and eternal essences of justice, knowledge, truth, and goodness. Socrates' efforts to articulate the inherent properties of universals are therefore not entirely unlike pre-Socratic attempts to discover the essential traits of natural phenomena.

Because types and forms can be realized in multiple instances or *tokens*, each different from the next, providing a full list of actual realizations does not capture a type's essence; to do so, scholars must discover the full set of its primary properties. That is what real (as opposed to ostensive) definitions do. Even after Plato's influence gave way to Aristotle's, and well before formulating natural laws became the central aim of science, classifying natural kinds such as types of rock (igneous, sedimentary, metamorphic) or biological *taxons* such as vertebrates by discovering their primary properties consumed most scientists' time.

Multiple realizability, also called *degeneracy* in biology, where it is ubiquitous, will play a key role in this story. The terms refer to the capacity of different events and processes to realize the same function or other higher-level property. The main idea presented in this introductory chapter is that according to the received understanding of types and kinds, individual tokens of a given kind were thought to differ only in their secondary properties. Their essential, primary properties, those that identify them as a type of thing—a given species, for example—are inherent, universal, and unchanging throughout. As late as the nineteenth century, this was even thought to be so in nature; Darwin's proposal that new species originated by evolving in response to selection was controversial for that very reason.

* * *

The way philosophers thought about essences and types evolved over the course of the history of ideas. As just mentioned, Plato concluded they were real transcendent forms. Aristotle, on the other hand, did not agree that form and matter could exist independently. In opposition to Plato's disembodied realm of forms, Aristotle focused on *embodied Substance*, the unity of *in-formed matter*.

Living things are quintessential examples of Aristotelian in-formed matter. In contrast to compacted masses of pebbles, for instance, organisms "hang together" in a unique way; they embody or realize a coherence that Aristotle postulated as the essence of substance.[4] Aristotelean substances are also *independent existents*, the other hallmark of reality for both philosophy and science.[5] On this view, actual lions are the coherent unity of the form Lion with the appropriate matter. In comparison to colors, sounds, or flavors, actual lions do not require observers to exist. In lieu of the otherworldly existence of Plato's forms, the essence of Aristotelean substances was therefore embodied self-sufficiency.

From both Plato and Aristotle's realist perspective, forms and universals are real. It was not until later in the history of Western thought that universals or types came to be considered mere labels or tools with which we think and organize our concepts and actions, or with which we design and conduct scientific experiments. This perspective, known as *nominalism*, held that the generality of common nouns is our doing; properties that define common nouns do not refer to objective essences. In contrast to realism, nominalism argued that all that really exists are actual particulars.

With the rise of scientific method in the sixteenth century, observation and analysis were elevated over a priori reasoning and synthesis.

In *Novum Organum* (*New Method*), English philosopher and political leader Francis Bacon proposed a new way of reasoning. In contrast to medieval Scholasticism's focus on a priori reasoning, Bacon's scientific method was based on induction, on inferring general conclusions from fine-grained, precise observations. In *Discourse on Method* a century later, French philosopher and mathematician René Descartes laid out a technique for knowledge acquisition that returned to deduction. Nevertheless, it required thinkers to begin with the most elementary and therefore indubitable ideas instead of accepted beliefs or revealed truths. In either case (Baconian induction or Cartesian deduction), the proper way to acquire knowledge was by starting from the elements (bare observations or clear and distinct ideas) and then proceeding to reconstruct the whole.

Threats to coherence are already present in this approach. Before Descartes, medieval thinkers like Christian Thomas Aquinas, Jewish Moses Maimonides, and Muslim Ibn Rushd (Averroes) followed Aristotle in the assumption that coherent substance—unity of form and matter—is the default basis of reality. What is the source of the coherence of embodied substance? Alas, Aristotle's writings did not fully resolve the question of how embodied substance, in-formed matter, is bound into coherent and unified wholes (Gill and Lennox 1994).

In contrast, influenced by modern atomists such as French thinker Pierre Gassendi and English experimentalist Robert Boyle, modern scientists and philosophers prioritized analysis over synthesis, parts over the whole. In doing so, the philosophical problem of identity was transformed from "How does essence in-form matter and confer on it its identity?" into "How do primary properties cohere into complex wholes?"

Whence coherence? Or, more precisely, whence the cause of coherence?

In Descartes' writings, Aristotelian embodied substance sundered into mind and matter, two distinct ontological domains, each with its own substance and each defined by its distinct primary property. Mind is defined by its primary property, res cogitans, whose essential trait is consciousness, and matter, res extensa, by its primary property, spatial extension. Known as dualism, the theory maintained that both substances, mind and matter, were real.[6] It is unsurprising, therefore, that by the middle of the seventeenth century, fears about whether coherence was even possible were being voiced. In his famous poem, *An Anatomy of the World* (1611), English poet and soldier John Donne lamented, "Tis all in pieces, all coherence gone."

But doubts about coherence did not begin in the seventeenth century. Philosophers as far back as Plato were sophisticated enough not to confuse form (an entity's core type identity) with shape. Justice was a Platonic form, for example, although it obviously has no shape. If forms are transcendent, however, how are actual entities related to the form that makes them what they are, that is, that imbues them with their essential nature?

Plato's famous allegory of the cave taught that physical objects are mere shadows of forms. Over the centuries, explaining how shadows or simulacra (material tokens) "partake in" or embody transcendent forms became a cottage industry. Plato thought of matter as a corrupting influence: when ideal forms get mixed in with matter, the result is a degraded reality. A few centuries later, and to avoid the Paulist and Augustinian disposition to label all matter corrupt after the Fall, Neo-Platonists such as Plotinus identified Platonic forms with God and suggested that they emanate into matter. Forms somehow ooze into matter?[7] Philosophy searched for a principle of coherence that binds transcendent forms to matter such as to generate coherent wholes.

The generation and preservation of coherence and its emergent properties is the central subject matter of this book.

Descartes' proposal that mind and body are two separate and distinct substances, each capable of independent existence and characterized by nonoverlapping essences, raised additional questions. Juarrero 1999 focused on how intentional causation and therefore purposive action (as opposed to reflex reaction) are even possible. How can two substances from entirely different ontological domains, the extended physical and the conscious nonphysical, *interact* at all? How can a nonphysical event like the intention to do X move the body such that the ensuing behavior carries out the content of that intention, doing X? Purposive action and the mind–body problem became the crucible of coherence. This book focuses on mereological causation, that is, on how interacting entities generate wholes with novel properties and how those wholes, once they coalesce, guide behavior. In particular, it focuses on the manner of causation that generates and preserves parts–whole and whole–parts coherence.

Interactions and Relations

Hypotheses about coherent wholeness and its primary and secondary properties intersect with views about the status of interactions and relations. As noted earlier, context and interactions, including fluctuations and perturbations, were presumed not to affect essential primary properties of fundamental elements or substances. To repeat, whether these consist of mass and atomic weight or thought and extension, primary properties were allegedly eternal, fixed, and universal, and they inhered in the entity whose essence and identity they underpin. Neither context nor history, not to mention process, alteration, or flux, could change those fundamental building blocks of reality.[8] "It's my nature." The essence of particles and substances (including minds) was internal, remained unaltered despite secondary modifications, and was capable of separate and independent existence. And, as always, relations and interactions with other particles or substances were considered neither essential nor foundational.

* * *

The field of chemistry is all about relations and transformations; chemists study interactions among molecules and charged particles. However, if relations and interactions are secondary, chemical compounds and chemical synthesis cannot then be fully real, *qua chemical*, as philosophers say. They must be nothing but aggregates of primary and essential properties of constituent physical elements. Chemical properties must be the accidental effluvia of clumped particles.

On this view, then, the apparent unity of chemical compounds, or living things, is illusory; wholes (including organisms) are mere agglomerations, the sum of their elementary parts. Material aggregates do not generate coherence; they are just masses of physical stuff. They are nothing but macroscopic clumps of elementary particles. They might appear to possess qualitatively novel capacities and properties like oxidation, life, sensation, and perception or even mind and consciousness. In reality, however, those are mere side effects—castoffs—of aggregated masses of primary properties. Those seemingly emergent properties and powers are ontologically derivative. They bring with them no powers of their own.

The principle that coherence is nothing but impotent aggregation is still presumed to apply generally, not only in chemistry. It motivates the philosophical perspective called reductionism, often characterized as "nothing but-ism." Reductionism maintains that because wholeness is nothing but particles and their secondary interactions, any seemingly novel properties of purportedly coherent wholes can be derived, in principle, if not in fact, from laws pertaining to its constituent elements. Colloquially, "Wholes are nothing but the sum of their parts." On this view, descriptions of *coherence* are either convenient labels or epistemic simplifications to which scientists resort because computing all the details of micro-interactions among primary properties takes too long.

In summary, reductionists maintain that, in principle, chemical, biological, psychological, sociocultural, or ecological processes, reactions, combinations, transformations, changes, and rearrangements of elements will all be fully accounted for by internal, context-independent, and fully quantifiable, essential—that is, primary—properties of fundamental particles.

The inevitable conclusion of the foregoing is that seemingly coherent entities like biological organisms and chemical compounds are causally powerless *as* biological and chemical wholes; if apparently coherent "wholes" are mere aggregations, their seemingly emergent properties are powerless. In the jargon of philosophy, coherence and emergent properties that characterize that apparent coherence are *epiphenomenal*, that is, causally impotent as such. The heavy causal lifting is always and only done by the fundamental elementary particles (van Gulick 2004). Accordingly, the arrow of explanation always looks downward, as the saying goes, because reality is grounded in the primary properties of elementary particles, which underwrite natural laws.

In short, misconceptions about causation cut across the previous considerations. According to the standard physics-based framework, as just described,

- Essences reside in internal and self-sufficient primary properties capable of independent existence. This implies that context and history are not fundamental features of reality.
- "Wholes" that appear to display novel and emergent properties are, at bottom, nothing but the sum of those primary properties and their interactions.

It follows that causal powers reside exclusively in fundamental particles and their primary properties. Bottom-up influences might bubble up from parts to aggregates, but with the following caveats: (1) any apparent coherence or wholeness is mere aggregation, and (2) seemingly emergent properties such as phenomenal awareness and feelings are, likewise, epiphenomenal characteristics of aggregated elementary particles.

Specifically, according to reductionism, our intuition that mental processes such as intentions and beliefs have powers to actively bring about meaningful, purposive actions is illusory. Thoughts, feelings, and intentions derive their powers and properties from biology; biology from those of chemistry; chemistry from physics. Properties that appear unique to biological organisms (such as being alive) or human beings (such as symbolic language) can, in principle, be inferred from chemical processes that constitute them. Causal powers that seem to issue from those higher-level properties can be derived from physical properties, at least in principle. It is not quite "turtles all the way down," however. The turtle at the bottom (at the level of elementary physics) is special. The primary properties of a-toms, reality's constituents (read now quarks and electrons), are the real and most simple stuff that does the causal work and provides explanatory power. Ultimate causes reside in and issue from there. Meaning and purpose are impotent.

Such is the dream of a theory of everything, the promise of an equation that spells out the lawful correlations among microdetails and from which everything else can be derived and precisely predicted. It is this intellectual heritage, I submit, that prompts well-known philosophers like Australian David Chalmers to espouse *double-aspect monism* or its close relative, *panpsychism*, the view that even the fundamental constituents of reality must have a built-in nonphysical aspect, mentality. Without literally building in consciousness into the most basic building blocks of reality from the start, hopes of showing how novel properties emerge from material stuff are doomed to failure because material stuff can only increase in quantity. The Cosmos does not actively create qualitative novelty. Full stop.

Efficient Causality

A particular understanding of cause-and-effect relations advanced by Newtonian science gave additional support to this worldview. By restricting cause–effect relations to *efficient cause*, the ontological status of coherent wholes became even more suspect.

I turn now to the backstory on causality.

The first systematic treatise of causation was Aristotle's *Physics*,[9] which argued that everything must be explained in terms of four Causes. The first three are Material cause, the physical stuff from which something is made; Final cause, the purpose or goal the substance brings about; and Formal cause, the inherent essence that makes it that kind of substance and no other. Consider a potter shaping a pitcher or a carpenter making a chair. Material causes of the pitcher and chair are clay and wood, respectively. Final causes are pouring liquids and sitting. (In nature, the *intrinsic teleology* or final cause of, say, acorns, is to become an oak. Teleology is built into nature.) Aristotle illustrated the fourth cause, Efficient cause, with descriptions of the potter's hands transforming clay into a pitcher and the carpenter's arm wielding saws or hammers[10] on wood. More recently, the preferred illustration of efficient causes is cue sticks striking billiard balls. Efficient causes are energetic forces acting on matter.

With the advent of modern science and philosophy, final and formal causes (which Bacon disparaged as "metaphysics") could be dismissed as "no part of science." Empiricism held that *material* cause could be accounted for through empirical observation of primary properties, and purposive or *final* cause came to be viewed as an irrelevant leftover from the discredited belief that the ultimate drive of all living things is toward perfection and God. The ghostly concept of formal cause became otiose as nominalism took hold.

As a result, modern science turned to Aristotle's *efficient cause*, the transfer of energy to matter with which forceful impacts bring about effects. After the sixteenth century, the term *causality* and its cognates came to mean exclusively efficient cause, the transfer of kinetic energy from an agent-cause to a body. Since the advent of Newtonian mechanics in the seventeenth century, cause–effect relations have been conceptualized solely in terms of motive forces. This perspective was amply rewarded with Newton's Laws of Motion and justifiably so. I hasten to state unequivocally that nothing in this work should be taken as a rejection of or argument against efficient causes or Newtonian science. When calculating planetary motions, however, adding a third planet yields a system that is chaotic and unpredictable. The laws of motion fail to predict their trajectories. Newton himself was aware of this so-called three body problem, an example of environmental influence completely transforming a system's dynamics. The puzzle was an early hint that context should not be so readily dismissed, but modern science and philosophy ignored its warning.

* * *

By definition, efficient causes exist separately from their effects. Relying exclusively on this form of causation, however, brought with it other, less well-known implications.

Aristotle's reasoning about efficient causes went as follows: since causes and their effects cannot simultaneously exist both before (as cause) and after (as effect of) themselves, causes are spatiotemporally "other than" their effects. The potter's actions start before they touch the clay. The motion of the cue stick begins before it hits the ball. The potter is spatiotemporally distinct from the pitcher; the cue stick is, likewise, "other than" the ball. The inevitable conclusion is that *self-cause* and other forms of "circular causality" are impossible because causes and effects would have to exist simultaneously: before (as agents) and after (as effects), a contradiction.

Conclusion: the view that reality rests on primary properties of elementary particles and that causality is exclusively efficient cause is at the root of modern science and philosophy's failure to explain the causal powers of systemic properties. Efficient causality cannot account for how parts become interwoven into interdependencies that bind together coherent wholes and persist despite turnover or deletion of component parts. Much less can it account for top-down causation from the emergent features of coherent wholes to those components or behavior.

Mereology

As noted, philosophers call relations from parts to wholes and wholes to parts *mereology*. According to the default ontological framework described (and that is still very much in place), parts cannot interact to produce wholes with strongly emergent properties—that is, with properties that are not reducible to the sum of the parts. Efficient causes produce mere aggregates. Emergent characteristics of wholes, as wholes, such as coordination and synchronization, cannot influence components because they could only do so as efficient causes, which is impossible since parts and wholes are not spatiotemporally distinct. And since aggregates are epiphenomenal, mereological "causation" from clumped particles to components is impossible (Campbell 1974; Kim 1989; Moreno and Mossio 2016; also see Murphy, Ellis, and O'Connor 2009). These misconceptions about mereological relations have occasioned no end of problems, many still unrecognized.

Working from the assumption that all cause is efficient cause, therefore, philosophers and scientists have resisted the idea of *recursive* or *circular causality* ever since. The problems of identity and intentional causation were perhaps the most insistent philosophical topics that wouldn't go away. Juarrero (1999) chronicled the epicyclic contortions to account for purposive behavior while refusing to accept recursive causality.

How can two separate and distinct substances from different realms interact in terms of efficient causality? How can mind activate matter? Descartes identified the pineal gland as the site where mental events such as intentions trigger purposive action. With the advent of thermodynamics in the nineteenth century, this maneuver ran into a different objection, this time involving modern science's conservation laws and principle of physical closure.

Physical Causal Closure and Overdetermination

Combine an exclusive reliance on efficient causality with the First Law of Thermodynamics, which holds that that the total amount of matter and energy in the universe is fixed. Matter can neither be created nor destroyed; it can be transformed (into energy) and back, but the total amount is always conserved. In combination, these two principles lead to a third, *causal closure of the physical*. This principle holds that (1) all states are physical states, and (2) all physical states are the effects of physical efficient causes. The unavoidable implication of these two premises is that proposing that the symbolic content of thoughts or beliefs can bring about changes in the physical realm violates conservation laws and causal closure.

The reasoning goes as follows: nonmaterial "causes" such as Aristotelian Formal and Final causes or Cartesian intentions would activate bodies by introducing themselves into the physical realm and activating them as efficient causes. In combination with the resolutely materialist presuppositions of the principle of physical closure, this insertion into the physical world would increase the universe's total matter and energy and violate conservation in the process. Conclusion: (allegedly) nonphysical thoughts, intentions, and other mental events as such cannot bring about physical effects without violating causal closure.

Purposive causation by nonphysical intentions would also imply an *overdetermined universe*, likewise prohibited by causal closure. The argument here goes as follows: causal closure requires that all events (including arms rising) be the effects of physical causes—that is, they must result from the energy transfer of efficient causes. But if intentions are Cartesian mental events and my arm rises because I intended to raise it, the resulting behavior is qualitatively different than if my arm rose because of an involuntary neuromuscular spasm or because someone pushed me. Intentional causation transforms behavior into purposive action, with all the legal and moral implications that entails (Murphy and Brown 2007; Juarrero 1999).

But if intentions as meaningful mental states can bring about purposive action, the arm motion would be caused by both my intention to raise my arm and the neuromuscular processes required by the principle of causal closure. Such double causation overdetermines the universe. Recall that on the standard understanding of efficient causes, causes and effects must be logically and spatiotemporally distinct, so on this account, intentions must be other than neurophysiological processes, which would also violate causal closure.

Conclusion: allegedly nonphysical mental events such as intentions, thoughts, and beliefs cannot cause actions as meaningful and purposive intentions to do X. These considerations reinforced the received view that systemic properties are epiphenomenal.

It follows from this reasoning that postulating top-down (downward) causation, from intentions in virtue of emergent mental properties down to behavior (Campbell 1974), violates conservation laws and the principle of the closure of the physical. Western philosophy's exclusive reliance on efficient cause and total dismissal of context bars naturalist accounts of mental states and mental causation at the root. By maintaining that identity rests on primary properties of fundamental material particles (while simultaneously excluding all forms of influence other than efficient causality and insisting that relations and interactions with context are secondary and epiphenomenal), understanding how mind and body could possibly constitute a coherent whole, much less how identity can be preserved and persist despite changes, becomes impossible.

As framed by standard philosophy of science, naturalism closed off debate about the causes of coherence and therefore about identity and individuation. Without a principle that explains the formation of coherent wholes from previously separate parts, natural science and philosophy of mind were also left without a way of accounting for either mereological relations or interactions with the environment and influence from the past. Nothing can generate outcomes that are qualitatively different than mere aggregation without violating causal closure. Reductionism followed. Human beings are nothing but atoms clumped together in time and space.

* * *

Over the centuries, several ways of resolving this impasse were proposed. The early twentieth century witnessed a flurry of philosophers who espoused an approach labeled emergentism.[11] In keeping with the reductionism required by natural philosophy, emergent properties were invariably deus ex machina contributions, literally injected from outside the physical realm into inchoate and passive matter. Emergentists such as French philosopher Henri Bergson were often labeled Vitalists in reference to the mysterious and externally provided "vital" force that they proposed animates living things. Natural scientists summarily ignored these proposals.

Other thinkers such as Scottish philosopher John Stuart Mill, Australian-born philosopher Samuel Alexander, French scientist Henri Poincaré, and Austrian biologist Ludwig von Bertalanffy cast about for a systems-theoretic approach to account for novel properties of wholes. Michael Polanyi tried to have it both ways by arguing that although ultimate boundary conditions were set from without, once in place, they can generate life without outside influence. Charles Sanders Peirce, a practicing chemist, knew there was something more to chemical compounds than mere aggregation (Juarrero and Rubino 2010), but in the end gave up on finding an ontological answer and, in keeping with Kant's epistemological turn, settled for a semiotic and "pragmaticist" account.

For the most part, however, the reductionist principle that aggregates are nothing but the sum of their parts and that seeming coherent wholes such as organisms like you and me are epiphenomenal, powerless to exercise causality as anything other than masses of atomic primary properties, ruled the day. One can sympathize with John Donne's lamentation that all coherence is gone.

But the problems of coherence and identity did not go away. Intuitively, it is difficult to shrug off the belief that there is some logic that holds organisms together over time despite perturbations. Organisms are organized; they present an internally coherent and persistent "hanging together" that is qualitatively different from mere aggregation. What makes coherent processes in general, including living things, cohere and persist as the same type of entity despite dramatic alterations? Suggesting, as early complexity theorists proposed in the mid-twentieth century (Prigogine and Stengers 1984; Maturana and Varela 1979; Pattee 1972a), that constrained and self-organizing dynamics might generate coordination and complex organization with qualitatively novel properties went completely against the grain of the received framework. With very few exceptions (Collier and Muller 1998; Cilliers 1998; Collier and Hooker 1999), mainstream philosophers of science disregarded complexity.

Meanwhile, the world moved on. Among the issues raised in identity politics today is the role contextual factors such as race, colonialism, class, culture, sexual orientation, and age play in personal identity. Environmental and historical influences are implicated in making all living things what and who they are. In biology, the new field of epigenetics has shown that the dichotomy between nature and nurture is too simplistic. The even more recent research on the metagenome in the gut microbiome, not to mention lichens and mycorrhizae, raises even further questions about symbiotic wholes. Are human beings more symbiotic wholes than individual organisms? Just asking this begs the question of this book's central theme: what causes coherence? In any case, the physics-based worldview from which our understanding of causes and effects, coherence, and identity derives continues to assign only secondary or accidental ontological status to context and history. It prioritizes separate and standalone existence over relations and interactions.

* * *

Mistrust in the dichotomy between essential and accidental properties surfaced in academic philosophy circles in the early twentieth century. Austrian philosopher Ludwig Wittgenstein and American philosopher Willard Van Orman Quine independently held that the distinction between primary and secondary properties was untenable. As an example, there are no necessary and sufficient primary properties that make all games *games*.[12] As students at Oxford University during World War II, Philippa Foot, Mary Midgely, Iris Murdoch, and G. E. M. Anscombe argued against the view that reality is wholly transcendent and that the faculty of moral intuition is independent of features in the world. These remarkable thinkers emphasized instead social, creative, spiritual, and biological forms of life as the source and weave of human life. Alas, the significance of their ideas is only now being recognized (Lipscomb 2021; Cumhaill and Wiseman 2022).

Since the received view of essences and type of entities was predicated on primary properties, the next philosophical step was usually a retreat to nominalism, the theory mentioned earlier to the effect that any pretense that types and kinds are real should be abandoned: common nouns like *triangle* or *scorpion* do not refer to real categories or to real entities marked by an internal principle of coherence; they are only linguistic units or epistemic cuts that frame our thoughts and investigations.

Remarkably, even though he was discussing the failure of essentialist definitions, the later Wittgenstein did not explicitly retreat to nominalism. He proposed instead that language captures "forms of life," a phrase often interpreted as the opening salvo of a social constructivist approach to language. Wittgenstein's phrase suggests to me, as it did to Anscombe, Foot, Midgely, and Murdoch, not an outright retreat from realism but an attempt to reimagine the world from the point of view of relations and circumstances, context and history.

An opening to a contextual and relational ontology, in other words.

Dismissing the influence of context is the original sin whose imprint continues to permeate many of these philosophical and scientific debates. This book will argue that recognizing that context changes everything reopens a path toward rehabilitating coherence, identity, and causation—from parts to coherent wholes and from emergent coherent wholes to parts, including intentional causation. It proposes a realistic and naturalist philosophy based on the principle that nature self-organizes and sorts itself into real but contextual *coherent dynamics*. Call its principle of coherence its *constraint regime*.

To be consistent, Platonic forms as well as final and formal causes must in consequence be reimagined in terms of interactional, extended, and *context-dependent interdependencies*. Thus understood, relational types are real and coherent patterns of energy flow, structure, and activities that form locally from contextually constrained interactions among individuals and that, in turn, as coherent dynamics, constrain the individuals and circumstances from which they emerge. Reimagining cause-and-effect relations, especially mereological relations between parts and wholes, and the influence of context and history on those relations, will be the hinge on which this reformulation turns. By taking context seriously in this fashion, coherence and wholeness, as well as meaning, "shall be returned."

This new interpretive framework makes room for a more expansive notion of identity and coherence based on interdependence and characterized by resilience and robustness. And generated by constraints.

Even John Donne might approve.

2

The Path Forward

Fortunately, the science of open systems far from equilibrium offers a new lens with which to rethink many of these issues. In particular, the dynamics of complex systems allow us to reframe our notions of cause and effect, wholeness, relations, context, and history. Ranging from tornadoes and chemical reactions to living things, so-called *dissipative structures* are open systems far from equilibrium that exchange matter, energy, and information with their environment. They self-organize and act as coherent totalities in response to constraints. They persist as themselves in a paradoxical state of dynamic stability despite being in non-thermal equilibrium. Borrowing concepts from nonlinear dynamics and complexity theory will help clear a science-grounded path along which our understanding of coherence and identity can be restored.

It is a trail that ventures into closed-off areas that the accepted framework barred philosophers and scientists from exploring. Rehabilitating the concept of coherence begins by taking context seriously.[1] By working backward from the failures described in the Backstory, the path to coherence requires restoring interactions and relations to their rightful place as coherence makers. It requires recognizing that context-dependent constraints induce integration and that the ensuing coordination leaves marks on the interactants. Notably, these marks can best be understood as multiscale and multidimensional *coherent dynamics*, patterns of energy, matter, and information flow that transform erstwhile separate and isolated elements into interdependent skeins. Having become interactants and relata, the erstwhile independent elements from which the coherent dynamics were generated are in-formed by the constrained relations in which they are now embedded and on which they are now conditional. These interlocking relations become governing constraints that hold those coherent dynamics together and contribute to their persistence.

By forging interdependencies in this fashion, embedding in context changes everything.

If not as efficient causes, by what manner of causality do interactions, relations, and context bring about change? The arguments marshalled in this book will conclude that coherence, wholeness, and identity arise from the operations of constraints. In particular, enabling constraints explain coherence formation and its emergent properties. These new coherences take the form of interactional types. Recursive, iterated, and multiscale constraints also account for self-organization, self-cause, self-maintenance, and self-determination (Etxeberria and Umerez 2013). Likewise, persistence, the temporal form of stability, also results from multiscale and multilevel constraints. This book will argue that because interactional types generated by constraints are multiply realizable, a novel form of mereological causation emerges: top-down causation as distributed control coded in analog form. These ideas encourage us to think identity anew, reimagined as resilient and robust—and causally effective—interdependencies generated and controlled by interlocking constraints.

I begin with a few intuitive examples of constraints.

* * *

In comparison with traffic lights and stop signs, traffic circles (roundabouts) are known to reduce the likelihood and severity of accidents, especially collisions with pedestrians and T-bone and head-on collisions. How do roundabouts bring about those effects? Not in virtue of Newtonian forceful impacts—that is what they are designed to prevent.

Roundabouts illustrate the workings of constraints. Consider how roundabouts alter pedestrian and driver behavior. Extensive studies of traffic circle design have shown that because traffic comes from only one direction, pedestrians keep track of oncoming vehicles more easily in a traffic circle. Drivers are also less confused than by the numerous intersections and stop-and-go of traffic lights. Roundabouts slow down traffic entering the circle (in contrast to traffic lights, where drivers commonly speed up at yellow lights). They even let drivers perform U-turns, which traffic signals often do not. None of these conditions is a forceful cause, and yet the constraints embodied in a roundabout's design reconfigure the behavior of people caught up in them.

Consider a second example of how constraints affect traffic flow. Traffic flow depends on vehicular interactions. As density reaches an instability threshold, the behavior of drivers begins to correlate and interdependencies among them appear. These interdependencies display novel properties. For example, if even a minor accident occurs, a stop-and-go pattern called "congestion shockwaves" forms from constraints contributed by the speed, flow, and density of vehicles on the road. Congestion shockwaves are population-level properties generated by constraints. They make visible novel—because relational—features that emerge from interdependent behavior. These characteristics are qualitatively different from those of the individual vehicles separately. For example, congestion shockwaves propagate backward (upstream) from the site of the collision even though none of the vehicles move backward.

Contextual constraints that intertwine individuals with one another generate coherent dynamics that influence those caught up in them. The workings of constraints explain how such coordination dynamics (Turvey 1990; Kelso 2009) emerge and then affect the constituents caught up in them; the dynamics are simultaneously constraining and constrained. In response to constraints, vehicular motions become correlated and interdependent, enacting the shockwave by moving in concert. The constraints that hold the shockwave coordination together ensure that the moving waveform persists long after the vehicles involved in the collision are towed away.

Constraints generating emerging traffic patterns are only one instance of constraints. The principles presented here are general. This book will argue that complex systems, living and nonliving, are coherent dynamics analogously generated by enabling constraints, as are convection cells like dust devils and hurricanes, as well as laser beams, synchronized pendulums, superconductivity, and superfluidity. In biology, symbiosis, homeostasis, and ecosystem stability describe such coordination dynamics. Social behavior and human cultures are exemplars of the workings of context-dependent constraints as well. Economic systems, also, are coherent wholes, interdependencies enabled by constrained interactions among producers, consumers, and traders of goods and services. These interdependencies, in turn (acting as governing constraints), modify the behavior of those caught up in that dynamic (Arthur, Beinhocker, and Stanger 2020). All are patterns of energy, matter, and information flow with emergent properties that result from interlocking enabling and governing constraints.

* * *

To repeat, complex systems are simultaneously constrained and constraining coordination dynamics. Operating on different materials under different circumstances and at different moments in time, each set of constraints underlies a distinctive variety of coherence with of its own qualitatively novel properties: convection cells rotate, laser beams cauterize, and homeostasis supports viability. Human beings reason and feel.

Coherent dynamics forged by constraints thus embody and enact distinct relational types of phenomena as constrained patterns of energy flow with emergent properties. Interactional types of this sort are realized and embodied in a variety of tokens or realizations, depending on local and current constraints. As an example, family relations differ in different cultures; each is a different token realization of family dynamics generated by different local constraints, in particular conditions, over time.

As envisioned here, unlike the universal and fixed essences of Aristotelian and Cartesian substances, coherence is knitted together by contextual constraints and is realized far from equilibrium as a flexible and dynamic equilibrium called *metastability*, of which bicycle riding and tightrope walking are examples. Metastable dynamics exhibit steady-state behavior that nevertheless remains in nonequilibrium throughout—thanks to constraints. As the traffic flow examples illustrate, qualitatively novel and context-dependent properties of interdependencies resulting from constraints are characterized by robust and resilient patterns of long-range and long-term coordination. These persist and do not quickly thermalize because, not despite of exchanges of matter, energy, and information with their environment and even while remaining in nonequilibrium throughout.

Critically, this book will argue that such coherent dynamics also have the power to exercise downward "causality," from wholes to parts. Interdependencies generated by constraints are not epiphenomenal; ask anyone caught up in one of those traffic congestion shockwaves. However, and this is the key point, although they influence their constituents top down, they do so as constraints, not efficient causes.

Cause and effect relations are therefore the central topic going forward. Catalysts and feedback loops, frameworks and scaffolds do not bring about effects as forceful causes. Nor do rules and regulations. As suggested above, some constraints enable phase transitions to coherent dynamics with emergent properties. But others lay down the coordinates of possibility space itself; they establish the boundaries of what is possible and what is not. Constraints such as the number of sides of a die, for example, limn the space of possibilities when throwing a pair of dice. Two six-sided dice thrown simultaneously have a possibility space of thirty-six possible outcomes. This boundary is determined by the number of sides on each die and the number of dice thrown. These factors are not forces; they are constraints. They have consequences as constraints, not as efficient causes. Analogously, direction, speed, and intersection design are constraints that change driver behavior, the care they take in navigating the roundabout, and so on.

Elements constrained to interact in certain ways produce qualitatively novel systemwide properties that loop down to affect those very parts. Pace modern science, the "causal" influence of constraints is corkscrew-like and mereological. Influence spirals from constrained interactions among individual drivers and pedestrians up to constrained and self-organized systemwide patterns of flow, and then back down, from systemwide dynamic down to the individual drivers and pedestrians whose constrained behaviors induced the macroscopic pattern in the first place.

Mereological influence is real and as true in economic systems as in traffic congestion. Adam Smith's so-called invisible hand in economics is one such simultaneously constrained and constraining dynamic. Constrained actions among producers, consumers, and traders create a new relationally defined possibility space that, as the systemic constraint we call the "invisible hand," simultaneously "governs" those individuals that populate that space. That does not make those individuals passive puppets; interactants are constraining as well as constrained. Complex dynamics like these are also inherently context and path dependent. Spatial configurations, timing, and even events that happened some time ago continue to make their influence felt in the present. For these dynamics, history and context matter. Chapter 5 will describe how this influence is also felt in the domain of public health.

Constraints are multifaceted; they operate, simultaneously, across different scales and dimensions. They can be enabling, governing, multiply realizable, and context dependent and independent. *Bottom-up, enabling, context-dependent constraints* such as positive feedback, catalysts and autocatalysts, recursion, and iteration take systems to a threshold of instability and precipitate phase transitions to novel and coherent interdependencies (to emergent and collective coordination dynamics).[2] Once those interdependencies have coalesced, and acting as *top-down governing constraints* through negative feedback, order parameters of that coordination dynamic preserve its coherence by stabilizing the system's individual components and behavior to within the boundaries of their interdependencies.[3]

As constraints not directly involved in energy transfer, governing constraints of coherent interdependencies can do all that without violating causal closure, conservation laws, or overdetermining the natural universe. Taking context seriously therefore requires us to take a different form of causality as seriously as we take forceful causes. Causality as constraint.

* * *

As noted earlier, taking context, time, and constraint seriously implies that coordination dynamics are novel and real types of entities that form in response to constraints. Such interactional types are multiscale and multidimensional ontic regimes of interlocking constraints that transform separate flows of energy, matter, and information into mutually dependent dynamic patterns; these are generic, temporal, local, and formed in response to contextual and historical constraints. Context-dependent interactional types and kinds are also *ontically indexical*; their emergent features refer to the contextual constraints they embody.

The constraint regimes of coherent, self-organizing, and autonomous types are realized, embodied, and enacted in individual tokens or realizations. In response to other, local, and timely constraints such as epigenetic influences, tokens of complex systems individuate over time. Components of a coherent dynamic do not blend or fuse into a higher-level monolithic entity; they retain their identity as a certain type of entity even as they individuate over time to become more distinctly themselves.

From this perspective, identity and individuation must be reframed as fluid yet metastable and persistent dynamic interdependencies forged by constraints in particular contexts and historical moments. The collective properties that irreversibly emerge from such interactions among individuals are inherently relational, reflecting the contextually constrained interactions that generated them. Implications for the identity of persons and nation-states alike are significant. Only isolated and closed entities are "independent"; in contrast, entities embedded in context constrain and are constrained by that context.[4] They individuate.

Different components, circumstances, and constraints leave their mark on the dynamic overall. In different contexts, including different histories or the interactants' psychosocial characteristics, different constraints—or the same constraints arranged differently—produce token behavior patterns and realizations with distinct characteristics and effects. Matter matters. Forceful impacts matter. But contextual constraints also matter. In mammals, neurotransmitters like dopamine and serotonin enable communication among neurons in the brain and possibly underpin awareness. But those same neurotransmitters infiltrate the gut, and there, dopamine protects against gastroduodenal ulcers. Different contexts or different constraints yield different results. Different paths specify distinct individuals, both synchronically and diachronically. An easy example: the exact shape and stability of actual snowflakes embody the conditions under which they were formed.

Contextual constraints integrate and organize all manner of energy, matter, and information flow. Constraints generate new coherent design patterns in the inanimate world as well as in living things—to wit, atmospheric dynamics, convection cells, and laser beams. Convection cells can appear in water, liquids, or gasses; they can be organized; and they can rotate clockwise or counterclockwise, depending on context. These coherent patterns of energy and matter can be enacted or realized in a variety of pathways and arrangements. They can be informational, structural, or dynamic. Collective and coordination dynamics generated by constraints thus bring the living and nonliving under one principle. Each can be realized in different material substrates and become differently specified depending on context and history. Because they are path dependent, each trajectory is unique. But all are the outcomes of constraints in open conditions far from equilibrium.

Consider homeostasis. It is not an organ; it is not a force. It is, ontically, a coordination pattern that arises from interlocking constrained processes among distinct biochemical, metabolic, neurological, and related systems. Homeostasis is a mutually constrained and constraining dynamic with the emergent function of maintaining the organism's metastability, which it carries out by regulating and controlling those component processes. As a result of that regulation and control, those individual processes in turn enact and realize homeostasis.

Phrased otherwise, enabling constraints generate the coherent dynamic that constitutes homeostasis by integrating and coordinating individual metabolic and other processes into a new *topology*, a new possibility space. Once set and operating as top-down constraints, novel order parameters of that new possibility space adjust and modify component entities and processes such that the organism remains "in the service of" a given range of biological viability and optimality. Health is an emergent normative metric against which those functional parameters are calibrated. The significant role of context in this entire process is evident.

Interactional types do not exist before interlocking constraints integrate diverse energy streams into coherent interdependencies. Types are not things. Whether it is a traffic roundabout, laser beams, the recently identified *integumentary system*, symbiosis, or homeostasis, coherent dynamics are mutually dependent relations generated by enabling context-dependent constraints. These interdependencies are real and powerful.

They consist of interlocking constraints whose dynamic interface with the environment filters, calibrates, and transduces separate and individual (incoming and outgoing) flows of energy, matter, and information, at various time scales and levels of organization, into a unified totality. This weaving, tuning, and translating enmeshes the various components into coherent flow patterns of matter, energy, and information. These realize a metastable and dynamic equilibrium, as established by a newly interwoven constraint regime. Once formed, that systemic governing constraint regime preserves its metastability far from thermal equilibrium such that the coherent dynamic persists longer than its individual components and tokens. General and real patterns of events, processes, and behavior are induced by contextually constrained energy flows, along with new order parameters and norms such as health, viability, metastability—and coherence and identity.

Predator–prey oscillatory patterns are also exemplars of interactional types. These cycles are metastable; they remain stable as a coordination dynamic whose constraint regime continues despite fluctuations in the number of individual organisms. Once formed as a metastable and persistent dynamic, the cyclic attractor of predator–prey dynamics constrains those individuals entrained into it such that the coherence of the overall dynamic is preserved and remains steady yet far from equilibrium. This it does by regulating and modulating the proportion of predators to prey such that it persists within a given range. And so, the cycle remains coherent and persists because its overarching interlocking constraints are invariant and metastable despite fluctuations in the numbers of the individual predators and prey.

A critical implication of this perspective is that multiple realizability is not disorder; it is evidence of the emergence of a complex form of order far from equilibrium. Such metastable patterns are interactional types as understood here, emergent contextually constrained and internally coherent patterns of energy flow. Coordination and synchronicity (Halpern 2020) are hallmarks of governing constraints at work. Phrased otherwise, governing constraints of roundabouts channel and select among possible driver and pedestrian behavior such that the organizing principle of the traffic circle is conserved. From superfluidity, oscillating pendulum clocks, or coordinated and synchronized marching, to biological function and even social organization, contextually organized, emergent macrodynamics are real and coherent patterns of energy flows brought into being by the workings of constraints. Once those coherent dynamics coalesce, they can possess unique powers and properties—to wit, the cauterizing power of lasers. Coherence brought about by constraints makes the difference.

The idea of context as used here spans time as well as space. Interactional types generated by contextual constraints lack the universality and fixity of Platonic forms and Aristotelian substances. Instead, thanks to their contextually constrained origins, actual realizations are tokens of a given type, tailored and cued to and by the spatiotemporal context in which, and from which, they form. Token realizations are specified according to the moment as well as the place in which the governing constraints of their type operate and to which they are uniquely responsive. Consequently, complex patterns of energy flow that are contextually formed and path dependent both integrate and process meaningful information and respond, as the occasion requires, *pros ton kairon* (Juarrero 1999).

Constrained and persistent interdependencies such as envisioned here constitute the ground of identity. The universe sorts and partitions itself by organizing contextually coherent interdependencies, increasingly extended in space and time. These coherent interdependencies are not epiphenomenal; they have ontic power to bring about effects in virtue of the interlocking constraints that generate and preserve them. Some of these constraints are inherited; others are enabled through decisions and actions. As mentioned, component energy flows are not fused or dissolved into larger wholes; encompassing constraint regimes are the outcomes of the enabling actions of individuals and their interactions. By entraining and synchronizing into interdependent totalities, behavior that is meaningful and appropriate becomes possible. None of the terms used here—partitioning, sorting, or hierarchy—should be understood to carry connotations of exclusivity and divisiveness. They represent both enabling and integrating as well as stabilizing and constitutive constraints.

Empirical evidence of ubiquitous multiple realizability generated by context-dependent constraints suggests that the cosmos possesses a deep capacity for integrating and thereby generalizing individual and separate energy flows into increasingly more encompassing, heterogenous, and differentiated context-dependent dynamics. As chapter 6 will describe, integration takes place in time as well as space. Preference for contextually constrained *diachronic* interdependencies—persistence—enables even more nuanced and context-dependent output modulation.

The three most significant signposts on the path forward are therefore the following:

1. Interactional type bring with them unexpected properties and powers; in particular, such interactional types are multiply realizable.

2. The multiple realizability of contextually constrained types underwrites the capacity for top-down control (from wholes to parts).
3. Multiple realizability also supports the evolution of analog control. As a result, the output (or behavior) of interactional types persists over longer periods of time and is particularly responsive and tailored to context. The rest of the book describes these dynamics.

* * *

Chapter 3 introduces the general notion of constraint. Chapter 4 explores context-independent constraints, those that take conditions away from equilibrium and lay out the canvas in which future complexification will occur. Chapter 5, "Why Context Matters," is a topical reminder of the role of context in epidemiology; it serves as a prelude to chapter 6, which is devoted to context-dependent constraints, those that take entities, events, and processes away from independence and link them into mutually interdependent coherent dynamics. These coherent dynamics constitute interactional, multiply realizable types.

Chapter 7 examines unique features of catalysts, autocatalysts, and feedback loops, three significant context-dependent constraints. Chapter 8 surveys a representative sample of other constraints, including "more-making," isolation, buffering, scaffolding, density, and saturation. Chapters 9 and 10 focus respectively on persistence and entrenchment, two important add-on constraints that preserve coherence, especially in human organizations. Chapter 11 describes Many-to-One (Patten and Auble 1980) and "Set-Superset" (Grobstein, in Pattee 1973) transitions as induced by constraint dynamics. This chapter also conjectures that multiply realizable coherent dynamics evolved a new form of constraint, *analog control.* Phase changes to coherent dynamics with analog decision-making and control subtend the potential for effective behavior, that is, actions that are nuanced and appropriate to context.

The intersection between hierarchy theory and coherence making is examined in chapter 12. Multiply realizable coherent dynamics are distributed and unranked *heterarchies*; they also often take the form of *holon*-like *holarchies* (Koestler 1967; Allen and Starr 1972; Salthe 1985, 2001, 2012) with rate- and scale-demarcated levels of organization relating to one another in terms of constraints, both bottom-up and top-down. Constraints operating between levels of such hierarchical organizations can influence processes and events despite not transferring kinetic energy; by operating as constraint, interlevel influences neither violate conservation laws or physical closure—nor postulate an overdetermined universe.

Part III examines these ideas as applied in contemporary philosophical and experimental work pertaining to these ideas. Chapter 13 updates the Backstory narrated in chapter 1, with special emphasis on the theories of functionalism and the 4E approach. Chapter 14 then examines philosophical arguments about the role of constraints in *supervenient* relations and their explanation. The chapter closes with a call for constraint-conditioned effective science—that is, for an expanded understanding of science that includes indexically defined principles that support counterfactuals—in specified contexts as enabled and governed by constraints. Delayed task experiments in neuroscience are examined in chapter 15. The studies described apply the statistical techniques of dimensionality reduction to tease out top-down control by collective dynamics. Results confirm the self-organization and in-formation (in neurological space) of dynamic attractors with emergent, task-defined properties capable of controlling behavior top down. Actions performed and controlled top-down in this manner will realize the emergent properties of those attractors. Chapter 16 offers concluding remarks.

* * *

This is the path to coherence that taking context seriously leads us on: it is a path that reinstates coherence and meaningful information as real and causally effective. It does so by rethinking essences and universal types, not as internal primary properties, or as eternal and fixed substances. Instead, interactional types as envisioned here are contextually induced patterns—and patterns of patterns of constrained interdependencies of energy flows. Because they lock in to real features of the world, contextual constraints that generate such patterns embody and enact real-world characteristics and interdependencies of the context in which they operate. Recursive iteration of feedback, for example, effectively imports real features of the context and thereby transforms the process into one that embodies and further specifies those features. Syntax and semantics become integrated as a dynamical coordination pattern.

It is important to reiterate that, in complexity theory, individuals matter; they are not pawns of collectives. Their actions matter; they are not simply reactions to external stimuli. Their behavior is in-formed by the self-organized dynamics that define them. That said, it is the workings of enabling and constitutive constraints—among individual entities, processes, and actions—that generate novel properties. Values, ethics, and morals can emerge over evolutionary time as effects of coherence making among interacting human beings (Artigiani 2021). The behavior of individuals entrained into those systems enacts and embodies those interlocking constraints that both simultaneously produce and constrain them. Had different individuals acted differently, the character of the interdependencies they generated would be different. That said, drawn into a different sort of context, each of us would be different, would behave differently. The coevolution of individual and context generally, including humans and their organizations, makes each journey unique; in the process, each is individuated, uniquely so.

* * *

This will be a work of speculative metaphysics. It presents an unabashedly ontological worldview by proposing that, in response to constraints, the cosmos partitions and sorts flows of matter, energy, and information into real coherent dynamics, and not just in living things. Ranging from laser beams and superconductivity to biological homeostasis, cognition, and even human symbolic thought, values, and the cultures in which they become manifest, coherent interdependencies are generated by enabling context-dependent constraints. These have real, effective, and emergent powers. Once formed, the interdependencies that enabling constraints weave together are governed by interlocking constitutive constraints that define and shape the domain—the possibility space—of possible realizers.

Objections of nominalists, positivists, social constructionists, and the rest are noted. However, so long as natural philosophers shy away from claims to infallibility and hold, with Nobel laureate Ilya Prigogine, that "nature is too rich to be described in only one language," synthetic speculation with a realist bent is justified if it blazes a trail for further investigation and research.

Context Changes Everything hopes to offer a fresh take on natural philosophy with an interpretive framework that reexamines how we think of relations and interactions with our world and with history. It especially aims to rethink concepts of cause and effect. Examining these ideas through the lens of complexity theory reveals the outlines of a synthetic and relational ontology with room for real coherence. Along the way, it pledges a renewed understanding of identity and individuation, as contextually embedded and coherent interconnectedness.

II

3

Constraints

An Introduction

Mechanical forces jostle things about but the motions they cause are Markovian. They forget their past; they are reversible. Likewise, Newtonian space and time are featureless containers that do not affect the basic properties of events that take place in them. In that received framework, measuring instruments with which physical processes are studied are assumed to be passive and transparent registers that do not fundamentally alter the entities and processes they record.

In the twentieth century, this pretty picture began to crack. Indubitable evidence that measuring quantum states necessarily alters the experimental outcome placed interactions, context, and time front and center.

Uncritically held assumptions about the role of context were first turned on their head by English physicist Thomas Young's famous 1801 double slit experiment that showed that light possesses a peculiar duality. Shining light through one slit onto a screen produced particle-like spots of light; shining light through two slits produced a wave-like interference pattern. What is light, really, a wave or a corpuscle? Far from revealing internal primary properties that remain unchanged regardless of the experimental setup, core properties of light differ depending on the interaction. Light appears one way in one context but different in another, depending on its interaction with the environment—in this case, a measuring apparatus. Instead of providing a faithful mirror of an essential and unchanging reality that is blind to interactions, the experimental context with which quantum states are measured alters results. Context leaks into the observed.

Over one hundred years later, German theoretical physicist Werner Heisenberg showed that one can know either an electron's exact position or momentum, but never both simultaneously. The information transmitted about the subatomic level is different depending on the experimental setup. Once again due to interference from the measuring instrument, Heisenberg's uncertainty principle set a limit to knowledge about subatomic particles.

The thought experiment that Nobel laureate Erwin Schrödinger described to Albert Einstein in a conversation was even more disconcerting. The uncertainty principle might have set limits to knowledge, but Schrödinger was concerned with its ontological implications. According to the standard Copenhagen interpretation adopted in light of the uncertainty principle, quantum systems should be conceived as clouds of possibilities, superpositions of states whose actualization depends on measurement. Schrödinger noted, however, that according to this interpretation, a real cat locked in a steel chamber and whose life or death depended on the state of a radioactive atom would be both alive and dead until a measurement is made.

Wave–particle duality was a dramatic wakeup call for physicists, but to the chagrin of working scientists since, context dependence has repeatedly been shown not to be restricted to the quantum realm. Because living things are quintessentially context dependent, biology was the discipline most recalcitrant to being shoehorned into a Newtonian mold. Organelles behave differently in a cell than isolated in a petri dish. Ontogenesis is heavily dependent on context. Epigenetic effects due to environmental interactions can alter gene expression and produce different phenotypic traits that persist even across several generations—without modifying DNA sequences. And we all behave differently around our parents than with our friends.

Water offers a surprising example of the propensity to ignore context. Although it is essential for life, water is often treated as inert background. Recent studies have shown, however, that, far from being the inert medium in which chemicals are dissolved, water is an active agent in at least 40 percent of the 6,500 known biochemical reactions. "Water exerted a huge influence on which chemicals survived and became a part of life, and which didn't. . . . The surviving molecules were the ones that were soluble in water. . . . 'That is how they were selected,'" Frenkel-Pinter states (Marshall 2021), implying a form of prebiotic selection by context.

Influences like these suggest that, lifted from the context in which they are naturally embedded, some events not only would not occur; they could not occur. Others would change dramatically. The keyword here is *embedded*, a condition unlike being plunked into Newtonian time and space or jostled about by mechanical forces. In various guises, context dependence is widespread. Nevertheless, settings, circumstances, conditions, and context in general continued to be ignored by the Academy. So-called special sciences such as psychology, sociology, and economics were even refused the label "science" precisely because of their context dependence.

* * *

To avoid becoming trapped in the historical baggage that the term *cause* inevitably brings with it, Juarrero (1999) relied on the concept of *constraint* to characterize the influence of context. As the next few chapters will describe, constraints are critical to coherence making. Autocatalysts and feedback loops, for example, produce long-range correlations and interweave patterns of matter and energy flow that display novel properties. And they do so as constraints, not mechanical impacts. This book will argue that interlocked interdependencies generated by constraints are the ground of coherent dynamics and their emergent properties. Moreover, constraints show that mereological powers and effects are real, bottom up, and top down, from parts to whole and wholes to parts.

This chapter introduces the general notion of constraint. Details of each variety of constraint are presented in the chapters that follow.

* * *

The late Canadian philosopher John Collier (2003a) identifies three prerequisites of complexity: a source of energy, *gradients*, and interactions that convert some of the energy influx made available by gradients to structure. ("Structure" includes "structures of process" [Earley 1981].)

Collier's first requirement is "a source of energy." It can be solar, geothermal, aeolic, hydraulic, chemical, biochemical, biomass, and so on. In the nineteenth century, the science of thermodynamics reintroduced the arrow of time, but it took the practical savvy of the Industrial Revolution to exploit constraints and extract work from energy flow for commercial purposes.

Harnessing energy is as important as the energy sources themselves; it allows energy to flow, and flow is necessary for order and structure to emerge. Collier's first requirement, a "source of energy," therefore implicates his second requirement, the presence of gradients, the first constraint.

In contrast to the concept of causation, the notion of gradients is unproblematic. The primordial gradient, cosmic expansion, originated with the Big Bang. On Earth, the ultimate source of energy gradients, of course, is the sun. Hydrothermal gradients near deep-sea vents are hypothesized to be sites where life on Earth might have originated. However, the full spectrum of energy is not equally available to all everywhere. Bats and dogs sense sound wavelengths that human beings cannot. Chemoautotrophs use energy from chemical compounds; photoautotrophs convert light energy to chemical energy. This major transition from anaerobic and extremophile microbes that rely on sulfate as their energy source to photosynthesizing cyanobacteria that produce oxygen was among the most significant of those innovations; had it not occurred, most plant and animal life would not exist.[1] The Industrial Revolution tapped fossil fuels; we do the same today with renewal energy sources. Microscopes, night scopes, telescopes, as well as computed tomography (CT) scans and magnetic resonance imaging, reveal energy spectra imperceptible to human senses directly. Different constraints that access different segments of the energy spectrum make a difference in the powers and properties they produce.

The second law is satisfied by the flow of energy along gradients. Transitions in cosmic, biological, and social evolution innovate by accessing new gradients, tapping new energy sources, and creating new information in the process. Since the early days of mechanical engineering, inclined planes are the textbook example of constraint. The steeper the slope, the greater their potential energy. Some natural gradients like cosmic expansion dissipate energy. Over time, the slope lessens and ultimately disappears. Cosmic expansion alone would quickly dissipate to thermal equilibrium without generating complexity. In contrast, nature's second gradient, gravity, is centripetal; it concentrates mass and energy. But gravity alone, too, would quickly implode in a massive black hole. By itself, neither cosmic expansion nor gravity produces coherence and coordination. Their coexistence throughout the cosmos, however, suggests the presence of other forms of constraints.

As understood here, then, gradients are only the first step toward coherence making and coordination dynamics. They are only one variety of constraints—*context-independent constraints*—to which the next chapter will be devoted. Other context-independent constraints such as buffers, isolation, and entrenchment will be explored in chapters 8–10.

Complexity formation therefore requires more than just a gradient; to evolve more complex dynamics, matter and energy must be coordinated and organized into coherent patterns (Bejan and Lorente 2004, 2008). It was only in the second half of the twentieth century with the advent of computer simulation (Abraham and Shaw 1992; Conway 1970) that science began to understand how coordination can tap and store energy in a way that isolated entities cannot (Nicolis and Prigogine 1977; Conrad 1972). Coordination harnesses gradients by capturing energy and converting it into persistent structure and order (Turvey 1990; Kelso 1995). Paradoxically, complex pattern formation in open systems far from equilibrium facilitates energy flow while simultaneously delaying heat death. Chapters 6–8 provide a detailed analysis of this process.

* * *

Collier's third prerequisite, interactions, can also be subsumed under the general idea of constraint. By interactions, complexity theorists mean relations other than the reversible bumping and jostling of Newtonian forceful impacts. Collier notes that self-organization is characterized by "unity relations," the information-carrying signature that signals a phase transition from unconstrained jostling to constrained coordination and order, such as the transformation from water vapor to the rigid structure of an ice crystal; from independent and separate photon streams into the coordinated alignment of laser beams; from algae and fungi to marvelous lichens; and from individual human beings to a distinct culture.

Constrained interactions leave a mark. They transform disparate *manys* into coherent and interdependent *Ones*. Constrained interactions, that is, irreversibly weave separate entities into emergent and meaningful coherent wholes. In doing so, they create and transmit novel information. That information is embodied in the coordination patterns formed by and embedded in context. Critically, coordination dynamics also leave imprints of the interaction on the interactants, which they change irreversibly. That imprint is new information (Shapere 1982).

The central question, of course, is What changes reversible bumping and jostling into *interactions that leave a mark—that create structure, order, and information*? The answer, just implied, is constraints, but in this case, a different type of constraint, context-dependent constraints (introduced in chapter 6).[2]

* * *

Coherence might be a better term for the unity relations of complex systems like snowflakes, tornadoes, lichens, living things, homeostasis, ecosystems, human practices, and cultures. Each of these is nothing other than a coordination pattern formed by different constraints with different stringencies, operating in different contexts. Critically, each of these overarching patterns is held together by a set of interlocking constraints; as a result, each displays qualitatively different emergent properties. Synchronization, coordination, and entrainment are instances of coherent organization, defined as a particular regime or logic of interlocking constraints. Coherence so understood marks a qualitatively novel form of organization and order that is absent in either isolated elements or clumped aggregates. Coherence is real; it is a relational and systemwide dynamic brought about by interdependence. All three of Collier's prerequisites therefore implicate constraints and constrained interactions.

The central insight of complexity theory is that constrained interactions among numerous variables in open systems in nonequilibrium can precipitate transitions to novel forms of order characterized by system-wide coherence, each with its own emergent properties. Coherence is not frozen and monolithic, block-like solidity. Complex order is a multiply realizable dynamic of constrained and constraining interactions. Elements in complex systems do not dissolve into a homogeneous medium or fuse into undifferentiated blocks. Rather, coherence consists of articulated and heterogenous interdependencies and covariances into which diverse elements are entrained and which now govern their behavior. Phrased otherwise, integration into coherent coordination dynamics is a form of generalization (Dean 2020, 36:40), a partitioning (Ladyman and Ross 2010) of reality into equivalence classes (Ellis 2016). Such coordination dynamics are contextually formed and multiply realizable.

Major transitions in evolution (Maynard Smith and Szathmary 1995) mark the generation of new coherences—that is, of new and overarching constitutive regimes that are the outcome of interlocking context-independent and context-dependent constraints. These contextually constrained interdependencies (Allen and Starr 1982) are best understood as *interactional types;* they enact covarying and interlocking constraints that embody qualitatively novel—because relational—information. Pointedly, as subsequent chapters will explain, because interactional types are formed through context-dependent processes, the information they carry and transmit is meaningful, not solely syntactical.

What are constraints and how do they generate coherent wholes? The next section delves more deeply into the general properties of constraints.

What Are Constraints?

Constraints are entities, processes, events, relations, or conditions that raise or lower barriers to energy flow without directly transferring kinetic energy.

Constraints bring about effects by making available, structuring, channeling, facilitating, or impeding energy flow. Gradients and polarities, for example, are constraints; others include catalysts and feedback loops, recursion, iteration, buffers, affordances, schedules, codes, rules and regulations, heuristics, conceptual frameworks, ethical values and cultural norms, scaffolds, isolation, sedimentation and entrenchment, and bias and noise, among many others.

Some of the consequences of constraints are *vectorial*. Gradients, such as those of electromagnetic fields and complex attractors, for example, guide energy flow in a particular direction. Computer settings, the interfaces of eardrums and cell membranes, and buffers and scaffolds all filter and select input such that energy flow is facilitated, directed, impeded, accelerated, harmonized, integrated, and so on, in one direction rather than another. Filters and interfaces that select inputs to an open system are also constraints. They do not directly transfer energy; instead they establish the context and conditions in virtue of which energy flow is possible, eased, accelerated, impeded, standardized, harmonized, channeled, directed, and otherwise influenced. Catalysts that promote or impede the integration of previously separate energy streams also affect the speed and direction of reactions. *Feedforward* and *feedback loops* use directionality to import the world into the very interdependencies and covarying relations they weave.

Spatial and Temporal Constraints

Constraints can coexist at a variety of scales and dimensions. They can be spatial or temporal, and both can simultaneously interact to influence how particular events change. Temporal and spatial constraints often "go without saying." And ignored.

Spatial configurational and design constraints are better known. The length of a playground's seesaw in relation to the height of its base establishes distinct background constraints on where the children must sit given their weights. These constraints must be simultaneously satisfied if the children are to see-saw at all. Spatial constraints turn the likelihood of *this entity here* conditional upon (and covarying with) *that entity there.* When constraints make place important, it becomes measured in terms of conditional probability.

Spatial constraints, that is, organize the world according to place or location. *Here*, *there*, *inside*, *out*, *up*, and *down* are all products of constraints that encode relational spatial arrangements and configuration as conditional probabilities. Embryogenesis is guided by spatial constraints. Given the location of a cell in a fertilized egg, it is more likely to develop into a muscle rather than a nerve cell. Organized spatial configuration is thus the outcome of constraint-induced symmetry breaks that configure reality into local, internally coherent and noncommutative types of entities governed by distinctive constraint regimes. Novel properties and powers emerge as a result.

These features are unexpected—downright bizarre, in fact—according to standard physics where, as noted earlier, time is reversible, and space is a featureless container. The features would be impossible if events are separate and do not interact.

In analogous fashion, principles of symmetry in Renaissance architecture stipulate that architectural features like turrets be placed in certain locations, but not in others, with some locations being more likely than others. Templates, blueprints, and scaffolds are other enabling constraints that assist in bringing about configurations or designs in space. Their constraints make possible or impossible, and direct, facilitate, or retard energy, matter, and information flow.

The term *spatial* will be used loosely throughout this book to include psychosociocultural situations such as economic conditions and social activities, as well as those physical, material, chemical, and biological conditions in which events and processes take place. A culture or community's traditions, as well as its morals and values, likewise specify activities that simultaneously constrain the attitudes and behaviors of members of that community. They specify where and how social practices must be performed.

Constraints that specify the exact sequence in which events must occur are temporal constraints. The term *temporal* can refer to interlocking constraints operating simultaneously at many time scales, cosmic, historical, evolutionary, or developmental. Temporal constraints make the likelihood of one event conditional upon one or more earlier ones. They include, for example, those interlocking constraints that span a species' lineage and an organism's genome and epigenetic profiles to guide development.

Adding temporal constraints to spatial ones immeasurably increases the potential for evolving greater complexity and an expanded capacity to evolve. And yet, despite their ubiquity in biology and human practice, temporal constraints in cosmology, physics, psychology, medicine, sociology, and ecology might be even less understood than spatial constraints. Circadian and seasonal cycles of constraints on metabolism and endocrine processes point to an important role for time in biology and medical interventions. Dismissed as unscientific not so long ago, however, chronobiology and chronopharmacology are only now being studied more rigorously.

Timing athletic training to estrogen and progesterone oscillations during the menstrual cycle, for example, appears to offer significant improvement in performance. Recently, oncology has also been studying the phases of tumors to determine if *chronochemotherapy*, timed in accordance with those phases or administered at a particular time of the day, might control tumor growth more effectively (Sancar and Van Gelder 2021). Bird, turtle, and salmon migration patterns are likewise calibrated to the progression of seasonal cues. Timing mismatches between annual migrations of birds and caribou on the one hand and on the other, those of plant sprouts or prey they feed on are examples of temporal constraints in action. As I write this, seventeen-year cicadas have come out of hibernation in certain regions of the United States. The constraints that enable and govern these unusual, thirteen- and seventeen-year cycles of the nineteen broods of periodic cicadas found only in North America are not yet fully understood.

Temporal constraints are at the heart of algorithms and protocols. Constraints themselves can concatenate: the activation of one constraint can become conditional upon the occurrence of an earlier one, for example. The general logic can be characterized as follows: given that X occurred, Y becomes necessary, impossible, or more or less likely. Given that X and Y have occurred in sequence, Z becomes overwhelmingly likely. Thinking of suffixes like -TION when playing hangman helps. Given -TIO, N becomes overwhelmingly likely.

As temporally constrained (not just haphazardly chunked) units, sequence of steps can bring about a degree of metastability that individual steps, performed alone or in a different order, cannot. Changing the order in which the steps of a recipe or an algorithm occur, or the duration of the interval between steps, changes the outcome. In consequence, entire sequences of events can themselves become organized units whose effects are qualitatively different from nonsequentially constrained events. The entire sequence itself changes the possibility space in which it occurs by becoming a constraint on subsequent events.

Childhood development is strongly influenced by temporally coded constraints. Studies of the effect of trauma in children indicate that there is a window of opportunity to correct negative consequences: those placed in healthy foster homes by age two showed cortisol levels like those of controls, but those placed in foster homes after age two produced less cortisol and showed a blunted stress response (*NewScientist*, February 22, 2020). (Normal myelination is likewise negatively affected by trauma that persists after age two.) As an example from a different domain, constraint regimes are also an important subject of research in clinical administration (Zhou et al. 2005), where hospitals specify postdischarge clinical instructions conditional on earlier sequences of inpatient events.

Temporal constraints are present in nonliving things as well. Surprisingly, sequential order that makes a difference has been found even at the quantum level: Heisenberg's matrices showed that different outcomes will occur depending on the order in which even quantum operators are implemented (Rovelli 2018; Barbour 2020; Halpern 2020).

In biology, time-dependent constraints support the evolution of taxa such as prokaryotes, archaea, and eukaryotes along temporally constrained as well as anatomical axes. Hominidae, *Homo erectus*, Denisovans,

Neanderthals, Cro-Magnons, and *Homo sapiens* become "phylogenetically historical individuals" differentiated by distinctive interlocking context-dependent constraints, especially temporal ones.

Spatial and temporal constraints intertwine. Some constraints are hybrid—that is, their influence spans both where and when. In the biological realm, the genetic code is the preeminent example of *spatiotemporal intertwining*. Gene expression and regulation of developmental stages are instances of spatiotemporal constraints at work. We noted earlier that cellular differentiation in early embryogenesis, for example, is conditional upon the cell's location in the fertilized egg. There is also recent evidence that phenotypic effects of genetic mutations are conditioned by earlier epistatic interactions, those where the effect of one gene is suppressed by the effects of a different already active gene. Since selection operates on phenotypic traits, we can conclude that history plays a significant role in the direction of evolution.

Even in the genome itself, messenger RNA (mRNA) molecules encode a set of instructions for how to assemble a protein by arranging amino acids in a certain order, given certain conditions. What mRNA encodes, if you will, is a schedule, a set of temporal constraints that establishes the order and context in which other cellular processes must assemble amino acids step by step. Critical to cell differentiation and ontogenesis in general, then, regulatory and modifier genes arrange and time the expression of structural genes, moment-to-moment, site-by-site

Analogously, cultural and religious constraints structure social space through prescriptions, proscriptions, taboos, and rituals. In human societies, many of these constraints involve food preparation and eating. Deuteronomy's proscriptions concerning preparing and eating meat and dairy products together might well have originated as health-related constraints. To ensure that they are strictly adhered to, constraints with potentially serious outcomes often become entrenched as religious commandments and sociocultural rituals. To remove cyanogenic content from the cassava root and make it safe to eat, it must be prepared and cooked in a precise sequence of steps. These prescriptions transformed into religious practice for the Indigenous Tukanoans in the Colombian Amazon (Henrich 2016). The relation of practice to food safety might be forgotten, but Tukanoans follow the proper sequence for cooking cassava . . . well, religiously. These examples of constraints highlight their power to effect change in believers. Constraints are not epiphenomenal.

Timing is a temporal constraint. Seesaws were mentioned earlier; now consider playground swings. No matter how hard a child kicks, or how often, to increase swing height, what matters is when the child kicks. If they get the timing wrong, the swing will not swing. In addition to the kinetics involved, much of playground swing dynamics is a matter of timing, being or acting at the right place at the right time. In another example, research shows that gymnasts performing a full circle on a high bar are more likely to complete the 360° loop the later they inject energy into the swing from shoulder, hip, or knee (Irwin et al. 2021). Jazz syncopation is an interesting case in point: is it a complex constraint of musical composition, or is it instead a violation of a temporal (rhythmic) constraint?

Timing is just as critical in medical care, where context dependence is everything. Failing to administer CPR in a timely fashion, or withholding orange juice from patients in a diabetic coma brings about death as surely as a lethal injection. If the orange juice is consumed too late, the diabetic patient might well survive but will remain in a vegetative state.

Navigating by sound would be impossible without precise timing. Coded in terms of temporal relations (the interval length and pitch between two sounds), the timing code of our hearing system allows us to determine not only where sounds are coming from but also how quickly the source will reach us. (The code is much more precise in owls, which rely on it as they swoop down to catch a scurrying mouse [Humphries 2021].)

Because of modern science's dismissal of temporal constraints, the classical Greek notion of *Kairos*, appropriate timing, has been lost to science. The plots of many Greek plays often turn on the violation of Kairos. In contrast, the Newtonian state of mind focuses exclusively on *Chronos*, the allegedly objective clock-time understanding of time. But, as the three-body problem of planetary motion mentioned earlier teaches us, thinking in terms of one size fits all laws can be . . . well, inappropriate. This is particularly true concerning actions pertaining to legal issues, morals, and medical interventions; unlike eclipses, these are eminently context dependent and must be performed *pros ton kairon* (as the occasion requires). Future chapters will argue that the flexibility to engage in appropriate (not hardwired) behavior at the right place and the right time evolved because it provides survival advantage.

As is well appreciated by ecologists and social scientists, the more contextually dependent the system, the less explanatory universal laws like Newton's will be. Developmental biology is acutely aware of the spatiotemporally constrained nature of ontogenesis. But, except for the role of *inertial frame* in relativity physics, contextual constraints have not been central to our conceptual framework about cause–effect relations.

Context has been marginalized from much of physics and philosophy. Pun fully intended, it is time to bring it back by making room for the idea of effective science, principles that support counterfactuals within specified spatiotemporal contexts (see chapter 14).

Once again, a central theme of this book is that constraints are not efficient causes. Timing the kick does not impart additional energy to the swing over and beyond the kick itself; neither does the exact spot where the child sits on the seesaw. Because it is not a thermodynamic process, timing is not subject to conservation laws. How constraints bring about effects therefore avoids charges of overdetermination and violation of physical closure.

* * *

By precipitating symmetry breaks and making entities and processes covary conditional on each other, constraints turn possibility spaces irregular. Constraints sculpt a rugged and multidimensional landscape, the possibility and probability space of what can happen at all, what will most likely happen, and, when it cannot, can or must happen. This possibility space can be conceived on the analogy of an epigenetic landscape (see chapter 4). The conditional probability of individual events changes because their constrained interactions deform possibility spaces and thereby bias what can and cannot happen next, what and when it must happen, and so on. Depending on their weight, children must sit at the right spot on the seesaw for it to seesaw, kicks on swings must be performed at the right moment, and a gymnast's timing means the difference between success and failure. The coordinates and boundaries of a constrained space are encoded as probabilities: event with probability 0 = beyond that possibility space.

Analogous to the way spacetime curves and becomes a gravitational influence, spatiotemporal constraints transform possibility space into a rumpled canvas that can subtend subjectivity and the view from here and now. When coordinates and topology deform in response to constraints, place and moment suddenly matter. The view from inside a trough is different from the view from a hilltop. By warping possibility space in this fashion, local spatiotemporal constraints therefore also underpin the emergence of perspective and subjectivity. The view from my here and now is not sub specie aeternitatis.

Because numerous constraints must be continuously satisfied on many dimensions and time scales simultaneously, possibility spaces also reconfigure moment by moment in response to those multiple constraints, entrenched as well as current new ones. Possibility spaces are thus defined by their probability contour (Buchler 1977) or dynamic signature (Kelso 1995). I call it its profile.

From the perspective presented in this book, then, possibility spaces are not solely epistemic; they are real, bounded, and sculpted by constraints. These can be physical, chemical, linguistic, axiological, psychological, sociocultural, ecological, and so on; all can coexist or not, depending on the demands of multiple constraint satisfaction. Constraints encode the statistics and the meaning of the world in which a given complex system is enmeshed (Humphries 2021). They render contextually constrained relations path dependent. As a result, history and context continue to shape today's possibility spaces, in biological lineages, ontogenetic development, cultural traditions, and individual actions. As an example, the fact that no new phyla have appeared since the Cambrian explosion suggests that boundaries and coordinates of biological possibility space became sedimented during that period; it suggests the presence of very entrenched constraints. On the other hand, the emergence of oxygen-producing bacteria changed the entire constraint regime of the planet; this shows that emergent entities can and do modify the canvas in which events play out as much as the landscape influences the individuals' opportunities. Complex systems are not puppets.

* * *

This chapter described constraints in general, setting aside for the moment the differences between context-independent and context-dependent constraints. The next chapter focuses on context-independent constraints. Context-dependent constraints such as catalysts and feedback will be in the subject of chapters 6 and 7.

4

Context-Independent Constraints

At thermal equilibrium, nothing is more probable than anything else; since there is no organization, neither is anything qualitatively different from anything else. Distinctions and thermal nonequilibrium—that is, inhomogeneities in energy flow—are prerequisites for coherence-making, information, communication, and work.

Context-independent constraints take conditions away from equilibrium.[1] They render conditions, events, and processes that were equally likely no longer *equiprobable* (Gatlin 1972). They establish the boundaries of uneven possibility landscapes (like fields) within which energy can flow and other constraints can emerge. Context-independent constraints turn the space of possibilities in which a system's events and processes play out nonuniform or inhomogeneous. They induce nonequilibrium. On a cosmic scale, gravitational and electromagnetic fields are examples of primordial context-independent constraints at work.

As mentioned, gradients of inclined planes are examples of constraints. Slopes of gradients in general, including those of cosmic expansion and gravity, operate as context-independent constraints. They establish nonuniform background conditions along which energy flows. Gradients in chemistry and biology are well understood. *Polarities* are constraints that organize a landscape such that energy flows only in certain directions. Anterior-posterior polarities in biological development, for example, represent inhomogeneities that, once in place, function as context-independent constraints that set the stage for subsequent processes. Sperm, eggs, and the epithelial cells of blastulas start out with built-in anterior-posterior polarity. Future steps in ontogenesis are strongly dependent on these preset inhomogeneities.[2]

Electrochemical gradients likewise establish the direction and speed of ionic flow across membranes. In a process known as chemotaxis, single-cell organisms orient their foraging according to concentration gradients of chemical substances. *Diffusion gradients*, in contrast, determine the rate at which molecules disperse through a medium. A lot of brain activity, too, is about ion diffusion into and out of neurons. In maintaining the correct *concentration gradient*, neurons consume a significant percentage of the brain's caloric use, pumping potassium into the cell body and keeping sodium out.

The point is a general one: conditions that promote or impede energy flow need not be material walls; they are inhomogeneities in possibility space. Inhomogeneity in possibility space is realized as different probability distributions[3] of entities, events, and actions within that space. Gradients and fields, for example, bias whether, how much, and in what direction energy is more or less likely to flow. Attractors described by the Lorenz butterfly equations likewise sculpt the dynamics of phenomena like weather patterns; they channel and direct processes caught up in them. Social organizations and cultures, too, are structured by analogous constraints that shape the practices and mores of their members. They strongly bias what their members will or will not do.

Context-independent constraints are the set designers of fundamental possibility space within which subsequent events can occur. Metaphorically speaking, if a theater's stage design has no door stage right, the chances of a director producing a play there whose script stipulates, "Exit, stage right," on that set are zero. Analogously, the cosmos's context-independent constraints shape the coordinates and initial conditions of cosmic possibility space; they initialize its prior probability distribution, so to speak,[4] the default likelihood of possible events. As the product of context-independent constraints, for example, the default probability space of a positively charged particle makes it highly unlikely that it will move toward the positive end of a battery or that unconstrained events will move from cold to hot regions of state space. The prior probability distribution of the possibility space ensures that. Discovering the context-independent constraints of a particular domain gives scientists a prior probability distribution on which to base a theory-informed scientific program.

Containers of Gas and Landscape Design

The operation of context-independent constraints can be illustrated with easily understood examples, the first a textbook example of gradients, the other two from landscape design.

Textbook examples of context-independent constraints include the following: consider a canister containing gas molecules at equilibrium with the environment, a state analogous to white noise. Not only does homogeneity convey no information, but it also has no identity, no character.

The container walls might as well not be there. There is no-thing there (Ladyman and Ross 2010). At equilibrium, energy flow is impossible, and work cannot be performed. Everything interesting, including identity and individuation, starts with inhomogeneity. A pendulum at rest is the equivalent of heat death.[5]

Inserting a partition into the container of gas at equilibrium and moving it from one end toward the opposite side creates an inhomogeneity. Gas molecules are now packed on one side of the partition; the rest of the container is a near vacuum. Gradients can also be produced by heating the gas inside the container. Doing so produces a gradient with the cooler conditions outside. Whether it is a board placed at an angle or a partitioned container of gas, these examples illustrate how context-independent constraints take actual conditions away from equilibrium. Planks angled at a slope or containers of gas at a higher temperature and pressure than the environment point to the presence of context-independent constraints; these change the topology of possibility space, from uniform and homogeneous to non-uniform and inhomogeneous.

* * *

Roman aqueduct engineers well understood the potential (in every sense of the word) of context-independent constraints. Constructing and maintaining the correct (1/1500) slope of aqueducts produced the disequilibrium that supplied the famed Roman *thermae* with a reliable flow of water. Aqueducts sculpt the possibility space along which flow can be channeled and for which they serve as context-independent constraint. They take conditions away from equiprobability; they are not forces that pump water through the channels.

The Renaissance gardens near Rome called Villa d'Este might be an even more impressive example of the deployment of context-independent constraints on a natural landscape. Built in the sixteenth century by superb hydraulic engineers, Villa d'Este has fifty-one fountains and nymphaeums, 398 spouts, 364 water jets, and 64 waterfalls, all supplied by over three-fourths of a kilometer of canals and channels, and relying entirely on gravity—not a pump in sight. The water ballets that delight visitors to this day depend on the context-independent constraints set by gravity on the terrain. Superimposed against that backdrop, the garden's network of channels and troughs then functions like the body's vascular system, a local context-independent constraint that regulates and directs the amount, direction, and manner of water flow that is even possible in certain locations but not others.

It took kinetic energy and efficient causes to construct Roman aqueducts and "bulldoze" the ground at Villa D'Este to create those gardens.

It also took energy to insert and move a partition inside the container of gas molecules. The point about possibility landscapes, however, is that, once conditions are away from equiprobability (and Renaissance engineers understood the nonequilibrium introduced by gravity), the boundaries and topology do not themselves transfer energy as kinetic forces (as would be the case if Villa D'Este's water flow were pump-generated like our blood flow, which is heart-pumped). Once the context-independent constraints of the natural terrain and the hydraulic engineering infrastructure are in place, the inhomogeneous topology serves as a background spatiotemporal constant that influences subsequent events. In similar fashion, once the body's vascular system is in place, the vesicles direct and channel blood flow but do not themselves directly transfer energy (Mossio 2013; Moreno and Mossio 2015; Montevil and Mossio 2015). Like traffic roundabouts, the design and condition of arteries and veins nevertheless have consequences. But their "causal power" operates as constraint, not efficient cause.

Context-independent constraints are configured into cellphone settings. Parental controls on mobile devices and their log-in credentials function in the same way. Those constraints set accessibility and timing requirements to which users' actions must conform. They determine which data are admissible, which outputs are possible, and which are not. Years ago, computer platform designers had a rough time figuring out how to program timing coordination to ensure proper coordination of data and analytics across software packages (Sheehy 2015). The backend development of IT platform interfaces sets up the context-independent constraints of the system's possibility space. Such preset inhomogeneities are important constraints often taken for granted as well.

Were there no constraints, there would be no system—only white noise. Cosmologically, constants such as the gravitational constant and Planck's constant are deep background constraints that determine the basic characteristics of the universe. Others include conservation and symmetry laws, Pauli's exclusion principle, and other fundamental context-independent constraints that sculpt the universe's basic inhomogeneities; they outline the boundaries and basic settings of its light cone. They determine the backdrop of constraints against which seesaws and playground swings can operate.

* * *

For living things, the edges and coordinates of the epigenetic landscape outlined in figure 4.1, together with the barely indented contour lines at the top (one wider and shallower hollow in the center flanked by low,

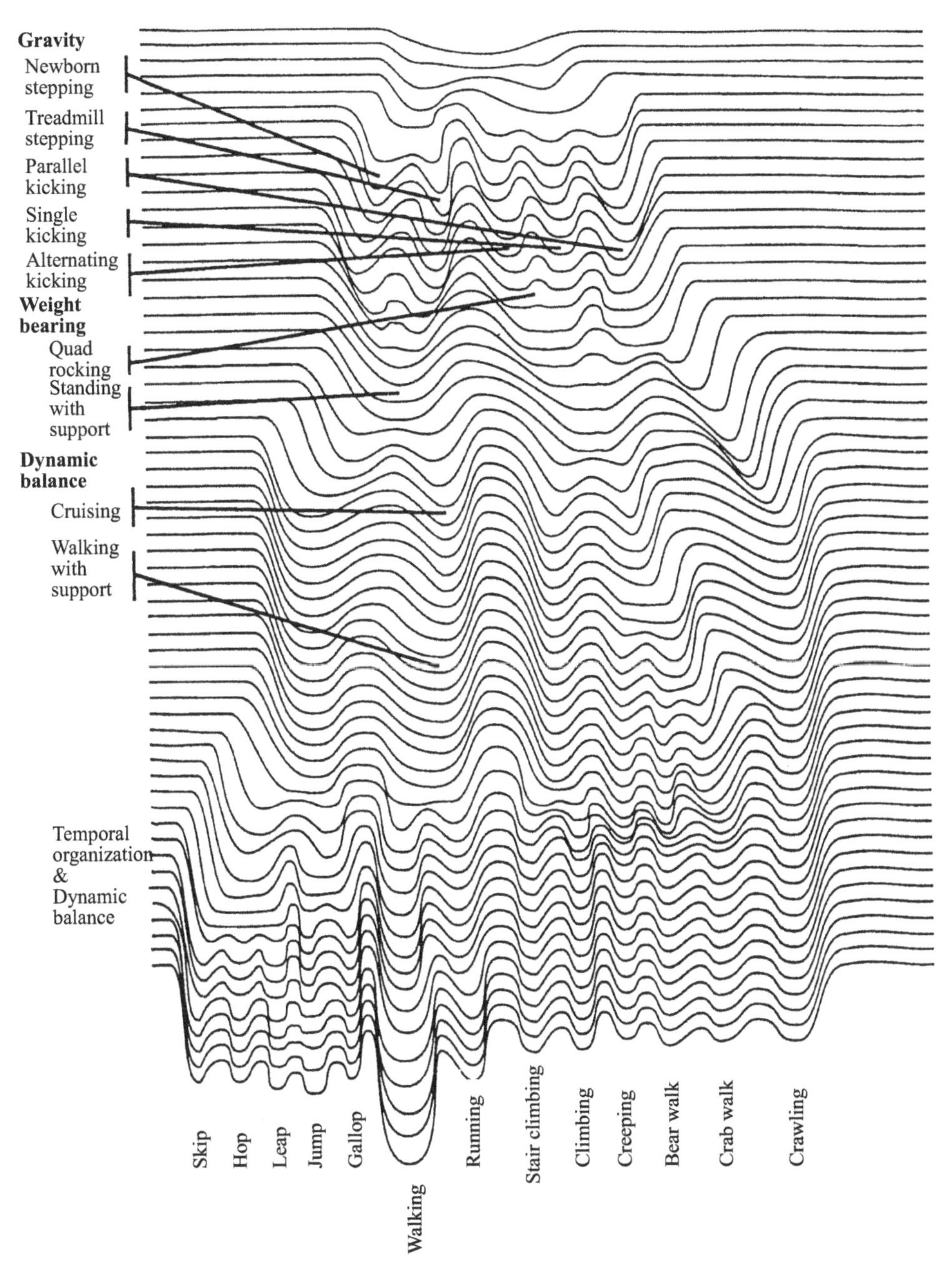

Figure 4.1
British developmental biologist Conrad Waddington (1905–1975), who first coined the terms *canalization* and *epigenetics*, was well known for creating illustrations of *epigenetic landscapes* to explain and visualize development (not evolution). In this example, the vertical axis represents (from top to bottom) the passage of time in the development of a fertilized egg. *Source:* Thelen and Smith 1994, 124.

mesa-like hillocks), represent an organism's inherited possibility space. For species, the constraints that define that landscape include constraints set much earlier by natural selection, as well as inherited epigenetic constraints just mentioned. They outline the axes and limits beyond which a phenotype's initially set traits cannot be modified by additional, contextual constraints; they determine the fundamental kind of thing it is (Grene 1974).

Examples of these fundamental constraints also include the sequentially ordered instructions encoded in its genome. These ensure the organism's unity of type is preserved. In evolutionary biology, *unity of type* (Brooks 2010; Agosta and Brooks 2020) describes the inherited constraints of a species' possibility space. Individual organisms stay true to type in virtue of their constraint regime—that is, the set of interlocking constraints they inherit and embody (Agosta and Brooks 2020; Brooks and Wiley [1988] call it H_{max}). Such lineage-defining constraints configure the regulatory settings that delimit the range of potential phenotypes available to a blastocyst. Those inherited constraints establish the lineage's context-independent constraints.

Waddington's illustration encourages us to interpret development and evolution in terms of constraints: hillocks are constraints that establish barriers to energy and information flow; hollows facilitate energy flow. Since energy flows from potential to kinetic, initial conditions in general (including at the birth of the universe itself) must be located on a hillock.[6] Thinking of initial conditions in terms of context-independent constraints can therefore account for boundaries of an embryo's multidimensional possibility space as well as its initial conditions (the slight indentations and hillocks at the top of the illustration)—both phylogenetically and ontogenetically. Local and progressively complex and differentiated troughs or chutes of probability are then sculpted over time by more local and context-dependent constraints over the lifetime of the organism.

Vague and Ambiguous Constraints

On the face of it, the growing inhomogeneities and ruggedness depicted in Waddington's landscapes appear to contradict the second law, which describes a tendency that goes from inhomogeneity to increasing homogeneity. The illustration above also suggests, however, that even initial fundamental constraints are not deterministic; they are *vague,* and they leave room for subsequent and more local and time-dependent contextual constraints to further complexify future foci of order and organization. In that respect, Waddington's two-dimensional illustrations are misleading.[7] Context-independent constraints are best conceived as preset configuration settings of the dimensions, possibilities, and limits of *multivariate* and *multidimensional* landscapes. These are probability landscapes whose boundaries and initial conditions represent context-independent constraints set by the organism's lineage, parental, and epigenetic inheritance, those that define it as the kind of system it is. The coordinates and initial conditions of these fitness landscapes represent the boundaries of epigenetic or developmental possibility space as well as its initial conditions (depicted in the warping in the first contour lines).

Significantly, Waddington's landscape also suggests that constraints responsible for those first shallow hollows and hillocks are multiply realizable. They bias but do not strictly determine[8] from the outset which of the multiply realizable alternatives within that possibility space will be actually realized. Note in figure 4.1 that the stable pattern cruising becomes subsequently specified in a variety of subdynamics.

Thinking back on our discussion of identity, it would be surprising if context-independent constraints were determinative. Strict one-to-one correspondence between unity of type and actual phenotype would leave no room for specification, individuation, or any sort of variation other than changes due to accidental mutations. It would certainly not allow the cosmos to evolve and complexify.

Instead, as understood here, the interplay between initial and subsequent constraints designs a sufficiently flexible "feasibility region" within which a variety of plotlines can play out. Those initial topological constraints differentiate later on into more specified pathways (deeper channels separated by steeper hillsides). Initial constraints are therefore vague;[9] their scope and reach are not entirely fixed.

Studies of the dynamics of cortical neurons confirm this approach. Excitatory input to cortical neurons is kept in balanced nonequilibrium by systematic inhibition, even across diverse conditions. That is, constraints that hold dynamic neurological equilibrium steady are adjustable, conditional on other counterbalancing constraints. For this persistent balancing act to be possible at all, innate and vague constraints must be *ambiguous*; that is, their governing constraints must permit multiple realizability. Such steady but flexible adjustments of neuronal excitation and inhibition enact a loosely balanced form of dynamic equilibrium (Ahmadian and Miller 2021). For such optimal balancing to be possible, a second variety of constraint, context-dependent constraints, will be necessary (see chapter 6).

This chapter closes with a few words on the built-in limitation of context-independent constraints that move systems to states of nonequilibrium.

Limited Message Variety

There is a principled limit to constraints that only take conditions away from randomness. Fields and gradients alone do not produce complex and persistent structures and dynamics. Likewise, more making and repetition (see chapter 8) are context-independent constraints that maintain or reinforce balanced but nonequilibrium conditions. But uncountably more identical tokens of one type is no variety at all. In the end, then, context-independent constraints play a foundational role, but the possibility space they establish severely restricts message variety. At the limit, gradients, regularity, repetition, replication, and other such context-independent constraints on their own reduce to one message (with probability 1). The possibility space would not become ruggedly complex and individuated with only context-independent constraints.

Whence variants of the same type of phenomenon? As proposed by the modern synthesis, only random mutations produce variants—but variants of what? Of a phenotypic trait that matters for survival. The operation of natural selection presupposes the presence of phenotypes that are tokens (variants) of a type of trait that matters for survival. However, random mutations alone cannot account for the unity of type of which the actual tokens' particular traits are variants (Brooks 2010; Agosta and Brooks 2020).

Whence coherence of type? The question is about ontology; their invariance and persistence over time make it highly doubtful that types are a mere epistemic construct.

Context-independent constraints may improve fidelity of transmission in communications systems but, alone, cannot transmit complex messages (Shannon and Weaver 1949). Nor can they underlie complexification. Standing and default background conditions set by context-independent constraints such as cosmic expansion and gravity take and keep conditions away from equilibrium. Gradients like cosmic expansion would end in a uniform and homogeneous heat death. Left to itself, gravity too would implode into a massive black hole. Same, same, same, more, more, more, again, again, again is unchanging information repeated endlessly. Massings, clumpings, and clusterings in gravitational wells—accumulations generally—cannot on their own become more complex. They cannot account for the persistent unity in multiplicity that tokens of coherent types embody and enact. The idea to be explored in the following chapters is that the universe's complexity is generated by more than one sort of constraint, operating concurrently and interacting together.

This second sort of constraint is context dependent.

Context-dependent constraints, on this account, operate against the backdrop of possibility space set by the primordial context-independent constraints of gravity and cosmic expansion, polarities, charge, and so on. *Context-dependent constraints* are defined as constraints that take particles of matter and streams of energy flow *away from independence* from one other. They weave together streams of matter and energy into the coherent and covarying pattern of a coordination dynamic. They make distinct entities and processes interdependent without fusing them into a monolithic entity.

Much of the rest of this book will be devoted to explaining how context-dependent constraints knit together multiply realizable patterns of flow into a complex tapestry of reality.

In contrast to context-independent constraints, context-dependent ones generate complex forms of coherence such as multiply realizable interactional types, degeneracy, pluripotency, individuation, and evolvability. Context-dependent constraints also underlie *metastability,* an emergent property of complex dynamics. That is, nonliving and living complex systems alike embody and enact a dynamic form of stability despite remaining in nonequilibrium. Such metastability, as future chapters will show, is an inherently flexible and adaptable but persistent form of balance; it also promotes resilience. The richly realizable and contextually constrained patterns of metastable energy flows cannot come about solely through context-independent constraints and mechanical forces.

It is tempting to conjecture that context-dependent constraints might have first appeared fortuitously in local gravitational wells at a critical density just before implosion into a black hole. Perhaps accelerated by density, the appearance of context-dependent constraints might have prevented implosion by irreversibly intertwining processes, events, and entities and thereby (meta)stabilizing them. Constraints of repulsion (electric fields with the same charge repel) and attraction might have rendered the constrained interactants relational. A recently discovered cosmic web of invisible dark matter seems to function as a reticulum that structures cosmic expansion and galaxy formation (Yang et al. 2020). Might this network be another (and in this case hybrid) fundamental cosmic constraint that keeps conditions away from equilibrium while at the same time generating certain interdependencies? Or might the Pauli exclusion principle be the primordial context-dependent constraint that keeps electrons in separate orbits and prevents them from collapsing into the lowest orbit? Critically, neither the web of dark matter nor the Pauli exclusion principle is a force; they are constraints.

* * *

Among the transformational changes wrought by context-dependent constraints is the emergence of *configurations*, patterns of coherence and dynamics in nonequilibrium whose metastability and resilience allow them to persist as themselves, adapt, and evolve. This book proposes that context-dependent constraints underlie identity and individuation, as interdependence. Chapter 6 delves deeper into context-dependent constraints.

Before that, however, chapter 5 provides an interlude that details the significance and complexity of context dependence in public health. In particular, it describes the nonlinearities and mereological relations of how infectious diseases spread, as well as the difficulties they pose to efforts to contain them.

5

Why Context Matters—An Interlude

Not all events are independent of context. The spread and severity of some diseases critically depend on the environment in which they unfold. As the COVID-19 pandemic taught the world, in such cases, context matters. We are learning that the same is true for climate change. Context-independent diseases might have given biomedical research its early easy pickings. As we discover that more diseases are context dependent, we will need to attend more to the role of context in medical issues.

In January 2020, early indications surfaced in social media that a new virus might be spreading in Wuhan province. Because of South Korea's previous experience with the MERS outbreak in 2015, the Korean population was favorably disposed to follow public health advice and prescriptions about this new threat. As a result, Korea experienced one of the lowest prevalence of COVID-19 in the world. In contrast, the kissing and hugging culture of Mediterranean countries significantly contributed to the spread of COVID-19 in Italy, Spain, and France, as did the widespread refusal to wear masks and respect social distancing in many parts of the world. Infections like COVID-19 and MERS are *context-dependent epidemics*. Such dependence on context is responsible for indirect health effects that can no longer be treated as irrelevant background. Contextual features that were previously taken for granted must urgently be foregrounded and addressed.

Science 2.0 and Medicine 2.0 Are All about Context Dependence

Individual cases of shingles are independent happenings; they are not dependent on context. Unlike COVID-19, someone's risk of contracting shingles does not depend on how many persons in the community are already infected. In epidemiology, *incidence* is defined as the likelihood that a given individual in a population will contract the disease. *Prevalence* refers to the proportion of persons in a population who have already contracted the disease. In the language of epidemiology, shingles is a disease in which incidence is independent of prevalence.

Shingles is context independent, in other words.

Binary classification into context dependent and context independent can be misleading. In the case of Lyme disease, for example, incidence may be independent of prevalence among humans. That is, individuals are not at higher risk of contracting Lyme just because they are around other people who have it. Whether others in the community have contracted Lyme is irrelevant to the incidence of infection. Nevertheless, incidence of Lyme is still very much dependent on other features—for example, on prevalence of ticks in that area in general—and on the prevalence of infected ticks in particular. In turn, the number of ticks is dependent on the number of deer and other hosts that are prevalent in that area . . . and so on.

Disease spread is not independent or dependent on context simply because incidence and prevalence are defined a certain way. Different answers to the same question, "What is my risk of contracting disease X?"—but asked about different diseases—indicate that independence or dependence on context is itself dependent on the scale and periodicity of that embedding context. It might be necessary to look further back in time and/or zoom out spatially to reveal the scale at which context dependence kicks in or washes out. Independence or dependence on contextual constraints at each of those scales and time frames, however, is real.

Context dependence is not subjective; it is objective, but relational—and induced by constraints. The uncritically held assumption that *relational* means subjective and therefore not fully real, as recounted in the first chapter, is an egregious legacy of Western philosophy's claim that reality resides in internal, primary properties.

In contrast to shingles, the spread and prevalence of many infectious diseases (such as COVID-19) are straightforwardly dependent upon conditions in the environment, the current phase of the outbreak's trajectory, and other similar factors. Context-dependent diseases were recognized early in the history of epidemiology.[1] As part of his 1916 study of the effects of quarantine on the transmission of malaria, British physician and Nobel laureate Sir Ronald Ross defined *dependent happenings* as those events where "the number affected per unit of time depends on the number already affected" (Ross 1916, 211). That is, an individual's likelihood of becoming infected or not depends both on what is going on around them and the history of the outbreak. Epidemiologists are interested in this problem because when incidence depends on prevalence as is the case in malaria and COVID-19, public health campaigns (including those pertaining to inoculations, social distancing, and mask use) become part of the context and can generate their own effects, direct and indirect.

Epidemiologists define *direct effects* as the probability that the intervention itself will produce a certain outcome—for example, the probability of becoming infected as an unintended reaction to inoculation. The reported effectiveness of a COVID-19 vaccine brand or therapy is about its direct effects. Since most vaccines have some side effects, these are counted as direct effects of the vaccine itself. Direct effects are not measured in relation to underlying or preexisting conditions. They are the effects only of the intervention itself.

Conditional direct effects, in contrast, are defined as direct effects of clinical interventions (such as ventilator use and antibody treatment) as conditioned upon the fact that, for example, patients receiving the intervention are already infected with, say, COVID-19. Conditional direct effects are calculated as the probability that ventilator use on COVID patients, for example, lengthens the time between infection and death. This context-dependent (conditional direct) effect became a critical issue early in the COVID-19 outbreak when ventilator use, which ordinarily helps patients suffering from respiratory illnesses such as the flu or pneumonia, was found to make matters much worse in COVID-19 patients.

The distinction between context dependent and context independent is a critical one for public health decision-makers because the effects they produce are often counterintuitive.

As noted, because contracting shingles is context independent, the likelihood of contracting shingles is not affected by the prevalence of shingles in the population. Public health campaigns about shingles therefore need not concern themselves with issues such as social distancing or the percentage of the population that receives the vaccine. That is clearly not the case for context-dependent diseases like COVID-19.

COVID-19 might have been a novel virus when it first broke out in early 2020, but public health agencies had dealt with context-dependent epidemics in the past. In the early decades of the twentieth century, yellow fever was endemic throughout the Caribbean and much of the United States. Sewers are ideal breeding grounds for mosquitoes that carry yellow fever. The more mosquitoes in a heavily populated area, the more people will be bitten and the faster the disease will spread. When incidence depends on prevalence, reducing prevalence—that is, changing the context—can contain an epidemic. U.S. Army physician Major Walter Reed undertook a massive intervention effort to successfully eradicate the prevalence of yellow fever in Cuba. Notably, for our purposes, it was not an inoculation campaign (there was no approved vaccine for yellow fever at the time). Intervention targeted instead those enabling conditions implicated in incidence and prevalence. Some enabling constraints were removed; others instituted. Massive cleanup of open sewers, an educational campaign to improve hygiene, and targeted fumigation effectively contained the out-of-control epidemic on the island in the first years of the twentieth century. U.S. Army physician William Gorgas implemented a similar program during the construction of the Panama Canal. Without it, the strategic passageway between the Atlantic and Pacific oceans would not have been completed.

That is, even without the benefits of a vaccine, the spread of yellow fever was halted by focusing on those socioeconomic constraints that enable contagion and spread of both the disease and the misinformation about it. Today, yellow fever no longer poses a significant threat.

In such cases, contextual constraints are far from epiphenomenal; they have significant effects. Consequently, epidemiologists and public health decision-makers must also examine indirect, total, and overall effects of public health interventions. These effects are not direct effects on patients; rather, they are the effects on individuals that occur because of the very success of public health campaigns themselves, from cleanup efforts to inoculation and social distancing campaigns.

The Indirect Effects of Context

As these examples show, the effectiveness of public health campaigns can have *indirect effects* on the prevalence of infection in the population at large. This is particularly true when the campaigns aim to change social behavior. The fact that context changes everything is most in evidence in indirect effects of context-dependent phenomena.

Indirect effects of massive intervention programs such as vaccination campaigns are measured as "the difference between the outcome in an individual [or a household] not receiving the intervention in a population *with* an intervention program, and the outcome in an individual [or a household] again not receiving the intervention—but this time in a comparable population with no intervention program" (Halloran and Struchiner 1991, 335). Indirect effects are calculated as probabilities conditioned upon the context (in this case, the intervention program itself)—for example, the percentage of a population that must receive the intervention for the indirect effect of herd immunity to kick in.

Effective public health campaigns indirectly but indubitably alter everyone's risk, even the risk to unvaccinated individuals. They do so by changing the landscape of the disease dynamics. In other words, context-dependent influences bubble up through population-level properties before looping back down to indirectly affect individuals. When interactions among individuals create a population-level emergent dynamic such as herd immunity, the entire profile of the outbreak suddenly becomes conditioned on this new property of the population. The likelihood that a given individual will contract the disease is now dependent, not only on their own proximity to others; it is also dependent on population-level features of the communities in which they live and work.

Free riders and coasters take advantage of such mereological and indirect effects. In populations where a widespread and effective immunization program reaches herd immunity, free riders really do get away with avoiding vaccination; their own risk of contracting the disease is dramatically lowered despite not using masks, practicing social distancing, or being inoculated themselves—because others did get inoculated, are masked, and do practice social distancing. To phrase it crudely, no individual is a percentage; herd immunity is a property of the population as a whole, a property that is measured in percentages. But that emergent, population-level property significantly mediates the way the disease plays out in individuals. Incidence suddenly depends on prevalence.

These are characteristically mereological (parts to whole and wholes to parts) dynamics. They are not confined to medicine; they are also at work in economic systems and convection cells, organisms and ecosystems. They are typical of complex, context-dependent dynamics, and their significant clinical effects should put an end to the controversy about whether systemwide properties are epiphenomenal or not. The Escher-like way in which parts-to-wholes and wholes-to-parts dynamics intertwine, however, makes them difficult to comprehend.

Indirect effects can be even more convoluted than the free rider example. Epidemiologists keep track of overall effects of massive intervention programs, which they define as "the difference in the outcome of an *average* individual in a population *with* an intervention program compared to the outcome of an *average* individual in a comparable population *without* an intervention program" (Halloran and Struchiner 1991, 333; emphasis added). Overall effects measure an intervention's effectiveness, on average. Effectiveness is an important population-level parameter for epidemiologists and public health decision-makers, not so much for individuals. Here is why.

Subtle differences in context (and how these are measured) can affect different individuals differently. In populations with large socioeconomic variations among subpopulations, the outcome of successful intervention campaigns, on average, can mask significant variations among those groups: one subpopulation might show limited prevalence while another shows a much larger prevalence.

Context-dependent dynamics are not one size fits all. As will become clearer in the discussion on hierarchy theory in chapter 12, because other (economic, social) constraints that are not directly related to the disease itself sort societies into qualitatively distinct subgroups, prevalence of context-dependent infectious diseases will vary depending on the circumstances of each subpopulation and on the individual's relation to that subpopulation. When context becomes part of the details of individual cases, comparing elderly residents of a care facility (who interact daily mostly with other elderly and weakened persons) with healthy twenty-year-olds is tantamount to comparing two different populations.

For example, as is typical of contextually constrained phenomena, subpopulations can interact in asymmetric ways. When some subpopulations provide most of the essential services to other subpopulations, the prevalence of disease in one or more groups (such as minorities, the aged, hypertensives, and so on) can asymmetrically alter the prevalence in another subpopulation (young, affluent professionals who can work and study from home). Because of these indirect and overall effects, sound bites that report the rate of COVID-19 without adding context and nuance are downright misleading. Information so generic tells individuals nothing much about their own particular risk. Despite coexisting in the same town, the two contexts are sufficiently different to constitute two different worlds. As we will see in subsequent chapters, they are two different attractors in a possibility space.

Communicating these subtle dependencies and contextualities is difficult. But because complexity and hierarchy theory are sciences that traffic in contextual constraints and not universals, they can reveal phenomena at much finer-grained resolution. Complexity science and hierarchy theory focus on subtle individuating differences, not on commonalities and averages. In consequence, understanding complex dynamics can facilitate context-sensitive and timely decision-making and action.

Finally, epidemiologists must study so-called *total effects*, a measure that combines direct and indirect effects of public intervention campaigns in context-dependent epidemics: this metric calculates "the effect of the intervention program generally combined with the effect of the intervention the individual in question personally receives" (Halloran and Struchiner 1991, 333). The total effects revealed can be downright perverse.

Consider the history of mass inoculations to prevent mumps and rubella in children a few decades back. The indubitable effectiveness of those campaigns produced dramatic positive and direct effects: many fewer children contracted mumps and rubella.

Counterintuitively, however, undesirable and unintended, but predictable, indirect effects also followed precisely because those public health campaigns were remarkably successful. Because mumps and rubella commonly spread more rapidly among schoolchildren, campaigns to eradicate mumps and rubella primarily targeted children for inoculation. Successfully eradicating these infections in children, however, had the following undesirable consequences: the average age of first infection (a population-level parameter) went up because those who did contract measles and rubella a few years later were (by then) older adults who had not been vaccinated years before. Some had not lived in that community during the inoculation campaign, or might have just missed the age cutoff at the time. Tragically, since mumps, rubella, and other childhood diseases typically produce more complications in adults than in children, as average age of first infection went up, morbidity among those infected also spiked dramatically (because they were contracting the disease later in life).

Other even more complex mereological relations show up in campaigns to contain context-dependent diseases.

Among the counterintuitive dynamics that epidemiologists and public health officials must plan for is the following: consider today's vaccines against diphtheria, tetanus, and pertussis (DTP). As is the case with COVID-19 vaccines as of this writing, a booster shot a few weeks or months after the first DTP shot is required because the first shot alone does not provide full protection. Direct effects, of course, are always higher if every individual in the population receives the booster as well as the first shot. That said, in populations where the initial intervention campaign was widespread and effective, subsequent exposure to relatively few free riders can, counterintuitively, provide the equivalent of a booster shot to those who did not receive the second shot—if exposure to the infected, unvaccinated person results in a mild reaction in the individual who only received the first shot. Enough free riders, on the other hand, can imperil the emergence of herd immunity. So, epidemiologists must even calculate the percentage of individuals in a population who are likely to be free riders and cheaters and compare it to the calculated percentage of the population required for the operation of herd immunity.

Indirect and total effects are not anomalies; they are real but top-down, mereological effects of a transformed collective dynamic (marked by a different periodicity and different parameters). It all depends on the role context plays in some disease dynamics. Explaining these kinds of dependent dynamics to the public is not an easy task.

* * *

How widespread are dependent happenings outside of infectious diseases? Recent research shows that individuals who have overweight friends and acquaintances are more likely to be overweight themselves—even when other considerations are taken into account. Obesity might be like some infectious diseases, where incidence is also affected by prevalence. "Social contagion means that if more people around you are obese, then that may increase your own chances of becoming obese. Subconsciously, we are affected by what people around us are doing. If you move to a community where a sedentary lifestyle is the norm, we tend to adopt that pattern as well" (Datar and Nicosia 2018, 240). The context dependence of socioeconomic characteristics nests the availability of food choices, which nests eating habits, which nests food preference, and so forth. Contexts and constraints intertwine along many dimensions and scales.

Scientists have been studying the transmission process of infectious diseases for over a century. Whether social contagion of obesity is merely a metaphor that arises because of correlations and not causation is still an open question. The dynamics of social transmission is a new area of research (Centola 2018). In biology, research in epigenetics is beginning to throw new light on the way molecular processes can be altered for generations by collective dynamics that implicate the environment and their effects inherited without causing genetic mutation.

Perhaps the tragedy of COVID-19, especially the role played by the spread of misinformation about mask usage, social distancing, and so on, will raise awareness of the fact that, for some health conditions, context changes everything. The next few chapters examine context-dependent constraints and how they both open new possibilities and close off others.

6

Context-Dependent Constraints

Chapter 4 described gradients and fields as constraints that take or keep conditions away from equiprobability; context-independent constraints like these make events in possibility space unequally likely. Unequal distribution in possibility space, as we saw, is a prerequisite of information and order. We turn now to *context-dependent constraints*, those that take conditions away from independence (Gatlin 1972). These weave together events and processes such that their interactions become mutually conditional on one another and on the history and context in which they formed. They now covary. The likelihood of events and processes influenced by contextual constraints is measured with conditional probability: X is more or less likely to happen, given the presence of constraint Y, in context Z. Chapter 5 used COVID-19 as a heuristic of the importance of this sort of constraint, especially in light of the way it generates mereological looping, from individuals to population, and back.

* * *

The following are three examples of the emergence of long-range correlations generated in virtue of context-dependent constraints. The first serves as a metaphor of phase transitions. The second illustrates interdependent dynamics among oscillators. The third is the textbook case of self-organizing, nonlinear, and far from equilibrium processes in the natural world. All three show how context-dependent constraints, operating against a backdrop established by context-independent constraints, weave global forms of order.

MacArthur fellow and theoretical biologist Stuart Kauffman[1] offered a clever illustration of phase transitions to coherence by randomly tying physical elements like buttons together, two by two. At a certain point, an abrupt phase transition happens. Well before most buttons are tied to another, the buttons lift as one when any one of them is picked up. Initially separate and independent buttons have become bound together in a two-dimensional mesh.[2]

Tying three buttons at a time or, in a simulation, on an average of 2.5 buttons, or 1.7 at a time, produces no such transition. In some cases, the system becomes chaotic; in others, the buttons just remain randomly connected. Two inputs per node, on the other hand, turns out to be a Goldilocks constraint; it is a context-dependent constraint that enables a phase transition to a new state with new properties. The coherent network is the interdependencies established by those enabling constraints. And networks are inherently relational entities.

Remarkably, the transition occurs when only about 50 percent of the buttons are connected. The abrupt transformation from separate and distinct buttons to an intertwined mesh represents a nonlinear phase transition from individual things to a two-dimensional, relational network. The architecture of the connectivity—what is connected to what, when, how, and by how much—suddenly becomes as important as the items connected. Context-dependent constraints induce relations that matter as much as the particles they link.

In contrast to context-independent constraints that take conditions away from equilibrium, context-dependent constraints therefore take conditions away from independence (Gatlin 1972) by intertwining isolated elements and processes; long-range correlations and interdependencies result. From these transitions there emerge novel, smeared out entities with different properties—in this case, a web. Enabling constraints, in short, transform separate entities into coordination networks characterized by relations and organization.

The second example is a textbook case. In the seventeenth century, Dutch physicist Christiaan Huygens discovered that pendulum clocks on a shelf interact through vibrational oscillations whose transmission is facilitated by the shelf on which the clocks sit. By adjusting the amplitude and speed of each bob, contextually constrained interactions transform separate oscillations of individual bobs into a coordinated unit. They synchronize into a common and periodic, harmonic wave pattern with a different frequency. Metronomes mounted on a common platform likewise entrain into a common wave pattern. These overarching wave patterns traced by coordination dynamics are longer and slower than the swings of the individual pendulums or metronomes. Different cycling rates at different scales of organization will play a key role in refashioning a new understanding of the boundaries of identity, as we will see later.

Convection patterns such as Bénard cells, dust devils, and tornadoes are the third example of the formation of interdependencies in response to constraints. Convection cells were among the first forms of dissipative structures studied (Bénard 1900; Rayleigh 1916). Consider a shallow pan of water or other viscous fluid at room temperature. At equilibrium water droplets move randomly,[3] each molecule identical to its neighbor, from which it differs only externally and numerically. If the pan is uniformly heated from below, as the bottom becomes hotter, a temperature gradient appears in the liquid. At a certain gradient, small convection streams form as the water molecules become increasingly correlated. At first, these fluctuations are damped by the overall system, but increasing the heat takes the system to a threshold of instability. Beyond a critical threshold, the context will amplify random convection streams and drive the system over a tipping point. As context-dependent constraints lock in, macroscopic molecular flows of 10^{20} water molecules moving in concert—acting coherently—suddenly appear. The system has abruptly switched from a regime of conduction to convection. As autocorrelated patterns of energy flow, these rolling columns of fluid break the symmetry of the pan of water. They are called Rayleigh–Bénard convection cells. The behavior of each molecule is now dependent on the presence and behavior of the surrounding ones.

Coordination dynamics like these (produced by context-dependent constraints on energy flows) form under far from equilibrium conditions. The initial links (such as a random convection stream) might be fortuitous. But if constrained interactions produce effects that are more metastable and increase energy flows overall, the context will amplify the dynamic and other couplings will no longer be fortuitous; they will become likely. The rolling hexagonal cells appear. The metastability of those cells itself becomes an encompassing constraint that self-reinforces. (See chapter 9 on stability as persistence.)

One can speculate that in the early moments of the cosmos, electrons packed in close proximity might have fortuitously and irreversibly become linked. Once connected, they displayed emergent and metastable geometrical properties, the linear, planar, and tetrahedral configurations of electron-pairs. Electron orbits of two nearby atoms, in turn, might have *entrained* in a constructive interference pattern, an even more metastable and self-reinforcing waveform we know as a molecular orbital. Just as water molecules behave differently in the context of a rolling Bénard column, atoms behave differently in the context of molecular orbitals. Context constrains. On the far side of each of these phase transitions, constituents are no longer separate and independent entities; they become relata and components of a new systemwide interdependence.

Coherence creation of this sort is best imagined as a "cusp catastrophe" bifurcation (Thom 1989), a phase transformation to a newly reconfigured topology—a new possibility space characterized by a new constraint architecture with new properties. Because the enabling constraints in question are context dependent, they lock in the resulting intertwined energy flows and information to real-world traits. In each case, the new context woven together by constrained interactions among individual elements constitutes an emergent set of now context-dependent and intertwined constraints. Because the enabling constraints in question are context-dependent, they lock in the resulting intertwined energy flows and information to real-world characteristics. Consequently, the closed set of constraints embodies, enacts, and transmits meaningful information about that world as a new possibility space whose components, properties, and behavior carry that information in their traits.[4] Interdependencies of this sort show that the cosmos satisfies the second law by generating coordination patterns and creating coherent dynamics that are metastable—thanks to the operation of constraints.

Coherence-making by constraints takes place in physical and biological complex systems small and large, from Bénard cells to human organizations and institutions, from family units to entire cultures. Entities and events in economic and ecosystems are defined by such covarying relations generated by enabling constraints. These mutually dependent relations are held together by interlocking constraints that can be called *constitutive-governing constraints*.

Adam Smith understood this bottom-up and top-down flow responsible for the interdependent (not merely aggregative) and persistent dynamic of economic systems. As mentioned in chapter 2, his famed "invisible hand" describes one such coordination dynamic induced by contextually constrained relations among individuals. Bottom-up, individual interactions intertwined by enabling constraints generate a coordination pattern of relational behavior. Once the population-level dynamic coalesces, the actions of individuals entrained in it change as a result of the resulting dynamic's constitutive constraints, as well as the history and circumstances that brought these about. Consequently, agents become consumers, producers, traders, and regulators; they act and respond differently to the embedding constitutive constraints in which they are caught up. We call the constitutive constraints that define this coordination its invisible hand.

Smith identified beneficial economic consequences of the invisible hand. Oligopolies, however, show that such interdependencies can also have a negative aspect: when product prices set by one firm are dependent on prices set by the other, and vice versa, the interdependencies of their constraint regime enact a loop that is difficult to break. Another notable example of the negative consequences of constitutive constraints is the depletion of public resources like the sheep-grazing area known as the "commons" or communal lobster-fishing areas off Rhode Island. In these last two cases, the dynamic generated by individually innocuous actions is a harmful collective property, the so-called tragedy of the commons. Failure to acknowledge mereological relations is one reason it took so long to recognize the real and powerful emergent properties of coordination dynamics.

Enabling and Constitutive/Governing Constraints

Keeping mereological relations of coordination dynamics straight is not always easy, as the COVID-19 loops showed.

Some context-dependent constraints are enabling constraints; others are constitutive (Mossio 2013; Moreno and Mossio 2016) or governing. *Enabling constraints* (Pattee 1973; Salthe 1985; Juarrero 1999) are context-dependent constraints that irreversibly link and couple previously separate and entities at the same scale as the constraints. By lowering barriers to energy, matter, and information flows such that independent entities become conditional on each other, enabling constraints drive parts-to-whole phase transitions to emergent coordination patterns, realized as mutually dependent relations.

Enabling constraints are nature's mechanism for coherence-making, generalization, and emergence. The rolling columns of fluid that constitute a Bénard cell are nothing other than interdependent, coherent dynamics generated by enabling constraints. Those coordinated and mutually dependent dynamics reflect the contextual constraints that induce and sustain them. For example, details like the speed and direction of flow will depend on the exact constraints under which each rolling cell formed. But coordinated and coherent dynamics have emergent properties their components severally do not, not least of which is their capacity to affect the properties and behaviors of those components that make them up. Phase locking, resonance, synchronization, and entrainment are emergent properties of coherently organized interdependent dynamics.

Enabling context-dependent constraints are therefore constraints that make the probability of one event conditional upon another. In phase transitions marked by bursts of entropy, they irreversibly generate emergent and coherent, metastable patterns of matter and energy flow. By *coherent* I mean that interlocking and covarying interdependencies brought about by enabling constraints hold systemwide patterns of mutual dependence together across spatiotemporal scales. They become autocorrelated, their interactions more tightly coupled, their feedback loops and relaxation times shorter, faster, and more regular than either with those of the environment within which they operate or with the elements of which the interactions are composed. The probability distribution, phases, frequencies, cycling rates, and timing of individual processes and events are woven together into multiscale, multidimensional tapestries of contextually embedded dynamics.[5] By *embedded* I mean that the system, including its components and its context, are intertwined in simultaneously constraining and constrained relations on which they mutually depend and with which they covary.

Complex systems ranging from convection cells to economic and ecosystems form and function in this fashion. As the pendulums, metronomes, and convection cell examples illustrated, the exact architecture of constraint regimes that manifest as waves or rolling columns keeps the particular interdependencies going. The pendulums and metronomes swing as one; the convection cells rotate as a unit. Complex entities formed by context-dependent constraints under conditions of nonequilibrium are coherent and persistent. Those interdependencies satisfy the second law: energy, matter, and information flow with greater ease as coordinated interdependencies than separately.

* * *

It is a central claim of this book that such coherent structures and dynamics constitute real and novel, interactional types of entities. Interactional types should not be reified; they are internally consistent, multiscale, mutual dependencies brought about by enabling constraints operating against a stable background set by context-independent constraints as described in chapter 3. They are measured with conditional probabilities. Critically, they are multiply realizable in distinct tokens. Not all constraints may be simultaneously satisfied, but those actual economies, or Bénard cells, or ecosystems that do exist satisfy numerous constraints simultaneously; they are the outcomes of *multiple constraint satisfaction*, a process of continuous adjustment of rates, weights, timing, and so on that satisfies as many constraints as possible.

Constraint satisfaction is an important form of "causality" that has been systematically ignored by modern science and philosophy. It can explain the generation and persistence of coherence. The effects of population-level properties on components, members, and stages of those interdependencies are also real. In response to multiple constraint satisfaction, components acquire new relational roles and properties—as members, periods, stages, and so on of a reconfigured dynamic with distinct population-level properties. As the previous chapter explained, a handful of unvaccinated individuals acquire new roles: they become shirkers and free riders because the likelihood that they will become infected suddenly drops to practically zero.

These terms capture real dynamics, they are not just convenient labels; they describe a real reconfigured probability distribution of events in possibility space. The probability (P_x) of individual event x abruptly transforms into a different probability, one conditioned on the full spatiotemporal context in which it is embedded (P_x/y, with y = embedding context). As members of a fully inoculated population, the risk of contagion for scofflaws drops precipitously because others have been vaccinated. The spatiotemporal context, what happened before and around them, has transformed the likelihood of what will happen to them next. Just as electrons possess different properties and behave differently in the context of a molecule than in the context of a single atom, the unvaccinated few acquire a relational property with respect to infectious disease dynamics: they are now *free riders*, a type of entity described by conditional probabilities. Disease incidence (the default likelihood P_x that a given individual might become infected) abruptly changed to a probability conditioned on the characteristics of the population in which they are embedded. Incidence suddenly switched to being dependent on prevalence.

Temporal Constraints

The spatial constraints of seesaws were mentioned earlier. In ontogenesis, too, spatial location of a pluripotent cell in the embryo biases the likelihood of whether it will differentiate into muscle or bone. But enabling constraints governing developmental processes are time based as well; temporal constraints of cellular differentiation and protein folding are as significant as the spatial context in which each step takes place.

Temporal enabling constraints are contextual constraints that turn entities interdependent in time; the second event becomes conditional upon the first. If X occurs, Y then becomes much more likely, or much more likely to happen next, or sooner rather than later, or faster rather than slower. Suddenly, X and Y are no longer separate and independent: Y is conditional on X; X has become a temporal constraint on Y.

Chapter 2 mentioned timing constraints in sports and medical contexts. It also noted that our sense of hearing works in terms of a temporal code. Physical and psychological milestones in child development are temporally organized; they are tuned to certain intervals and ordered sequences. Timing is implemented by temporal constraints that change the likelihood of future processes and events. Rules and regulations, codes and protocols, and norms and ethical principles also commonly operate as temporal constraints; they are instructions or procedures that specify a precise temporal order in which to do what, when—as well as where.

Often operating across several time scales simultaneously, temporal constraints bind together previously separate entities into complex diachronic interdependencies with emergent properties, coherent sequences in which what happened before biases what can and does happen when and next. As the sequence of cassava root preparation mentioned earlier illustrated, if a series of steps leads to greater stability, sequences as coherent units can form and be selected and repeated (Henrich 2016). The sequence as a unit reinforces because its order generates outputs that are more stable: it is why becoming an emergency medical technician is often a prerequisite for becoming a firefighter.

Another food preparation example (Henrich 2016) is illustrative. Without undergoing a particular sequence of processing steps, nardoo seed–like spores are indigestible and contain an enzyme, thiaminase, that inactivates vitamin B1.[6] Properly detoxified, however, nardoo is a main source of nutrients for Aboriginal hunter-gatherers in Australia. Performed in order, each step in the nardoo preparation process increases digestibility by breaking down thiaminase. The steps concatenate into a particular sequence which, as a unit, brings metastable results. Performed in order, the entire food preparation process brings into being a social practice, a population-level emergent property organized in time that yields adequate nutrition and is more metastable than had that exact series of steps not evolved into a coherent unit (Henrich 2016).

The full sequence itself, that is, becomes chunked into a context-independent constraint, a ritual or protocol of temporally coded segments. Temporally constrained sequences do not just start and stop; they do not track "one damn thing after another." They are now organized into episodes, stages, or epochs, with beginnings, middles, and endings. Even toddlers understand the difference between a parent who just stops cold while reading a bedtime story, as opposed to one who stops reading at a logical resting point in the story. Each segment embodies and enacts an internally consistent logic that holds a set of conditional dependencies together as a diachronic unit, a coherent, coordination dynamic organized in time, and characterized by emergent properties. That emergent property is a meaningful narrative.

Complex systems are therefore historical, not merely temporal; they embody temporal constraints in their very logic. They carry their history on their backs, as it were. Mutual relations of temporal constraint intertwine individual cells, tissues, organs and their processes into path-dependent complex systems. Self-reinforcement due to combined temporal enabling constraints is also path-dependent and historically constrained. On top of satisfying inherited context-independent constraints, embryological and child-development processes are also influenced by other more local and immediate spatiotemporally dependent constraints, each of which becomes active and is reinforced depending on timing and context. Changes that mark the onset of puberty must be held at bay during childhood, for example, and livers remember years of alcohol abuse. Historicity emerges with the formation of complex systems.

Cardinality and Indexicality, Ordinality and Placement

We noted that some constraints (as in gravity) can lead to clumping. Clumping makes for cardinal measures, but not complexity.[7] Prior to the introduction of context-dependent constraints, *cardinality* (quantity) was the only metric available. The clumps grow in size. More and less, bigger and smaller are about increasing magnitude, mass, or agglomeration, which are effects of context-independent constraints such as gradients. In a linear world, no qualitative novelty would arise.

Spatial and temporal constraints, in contrast, produce *indexical ordering*. Indexically defined relations explicitly refer to the spatiotemporal context that generated them and the emergent properties that govern them (discussed further in chapters 11 and 12). *Ordinality*, the position of an event in a sequence, is an indexical property that emerges when temporal, context-dependent constraints weave together a set of local interdependencies in time. The various conditional probabilities demarcate its spatiotemporal order. *First*, *second*, and *third*, or *before* and *after*, are emergent ordinal properties of points in phase space structured by temporally organized constraints. Significantly, this distinction is possible only because individual steps in temporally constrained sequences are not independent of each other. The burst of entropy with which irreversible phase transitions are paid marks a qualitative transition to a now temporally organized *phase space*. This new space represents a novel four-dimensional landscape, a new distinct constraint regime organized by time as well as space.

To recapitulate: classical thermodynamics demonstrated that temporal order is real. However, according to the received view of physics, micro-events forget their origins and the path they traversed. Chapter 9 will argue that there is no continuity proper in classical thermodynamics. In contrast, context-dependent constraints in open nonequilibrium conditions link separate events by making them conditional on one another. Time as well as space is incorporated in those enabling constraints that generate coherence-making and coherent sequences. Timing as well as place or situatedness becomes important. When context-dependent constraints are in place, that is, organization matters, temporal as well as spatial. The step-by-step instructions of genes, procedures, protocols, recipes, and musical scores describe such temporal constraints.

Contextually constrained coordination dynamics as understood in complexity theory therefore carry the imprint of their past and their environment in their present structure and dynamics. They embody and reflect the trajectory they have traversed. Temporal constraints weave together coherent sequences into extended narrative plotlines, biological lineages, and the life spans of individual organisms as well as cultures, all the while remaining in nonequilibrium. Depending on the details of how those constraints are implemented locally, these temporally coordinated sequences are actualized as progressively individuated systems that, also thanks to path dependence, blaze unique trajectories whose dynamics reflect that history. Spatiotemporal constraints thus turn entities and trajectories qualitatively distinct. To repeat, they are, for that reason, not only temporal; they are historical. Their past is etched into their present; their constraint dynamic leaves a mark on those they influence.

As examples: even raindrops, sleet, hailstones, powdered snow, and snow crystals are constrained trajectories that "condensated" (Burgauer 2022a) into distinct physical structures. Their emergent characteristics make manifest the context and the moment in which they were created. To illustrate, graphite and diamonds are made of the same carbon atoms, but the different constraint architectures in which their atoms are arranged actualize, in their very structure, the different atmospheric conditions such as temperature and pressure under which they were created. In consequence, their qualities are different.

Long-lasting temporal constraints and the long-range temporal dependencies they induce also underpin *persistent coherence*, the temporal counterpart of stability (Pascal and Pross 2015; see chapter 9). Persistent coherence is apparent in musical scores. Recurring musical refrains tie together longer segments of music. No longer independent notes, they are now components of a more encompassing unit, a musical phrase, a cadence, or an arpeggio. These sequences are temporal interdependencies whose components constrain and are constrained by the longer-term correlations in which they are embedded and that persist throughout.

Melodies are therefore path-dependent temporal wholes. Just as heritable constraints shape biological lineages over a longer time span than that of individual organisms, note order and tempo create melodies. Composers select notes for a musical score in light of those that came before and the overall composition in which those notes will appear. As an indication of the holistic yet differentiated nature of harmonies and melodies, expert musicians, for example, can hear other notes of a harmonic chord and even anticipate the next notes and chords in a cadence. Not only do individual notes reverberate as a chord's harmonics, but each note also continues to reverberate and resonate while the next notes of an arpeggio are played. Resonance can even expand. During a performance, brainwaves of members of the audience synchronize with those of the performers (Martone 2020). Such effects appear on the far side of a phase change that has dramatically transformed two independent possibility spaces into a more complex and spatiotemporally extended constraint regime.

* * *

These examples make clear that enabling constraints generate coherent dynamics whose boundaries need not be tangible, material structures. Rate differences between the circular flow of warm air constrained by low pressure at the center of the vortex and its outer edge mark the boundaries of a tornado. (Recall that differences in pressure are constraints that demarcate gradients.) Rate differences between full predator–prey cycles and the life span of the individual animals entrained in those cycles constitute the boundaries between parts and wholes. When enabling constraints weave together new coherences, autocorrelated dynamics are "lifted"—differentiated—from the contextual backdrop from which they emerged. Hurricanes, for example, are coherent structures sufficiently distinct from their environment to be visible from space This notion of identity is quite different from the inherent essential traits of Aristotelian and Cartesian substances. It is grounded in persistent, extended, and dynamic interdependencies among individual entities; between entities and processes on the one hand and conditions in the environment on the other; and between all of these and the past.

Whenever such interdependencies coalesce, individual components become *relata* aligned in the service of the more encompassing whole.

This is not to diminish the significance of efficient causes (such as the push that starts clock pendulums swinging in the first place). The point of these examples is to highlight conditions that serve as context-dependent constraints and take previously separate entities, be they atoms or clocks, away from independence (Gatlin 1972). The shelf on which the clocks sit, for example, does not add energy; it facilitates the flow of oscillations. Coherence-making and complexity formation at every scale are generated by a combination of both context-independent and context-dependent constraints. By irreversibly linking the probabilities of their occurrence, context-dependent constraints transform individual oscillations into interdependent components of new and overarching type of entity—molecules or waves. Without the shelf, there could be no synchronization. Without the characteristics of electron orbits and their proximity to other atoms, molecules would not form. Once formed, however, new constraint architectures such as waveforms, molecules, and population dynamics shape new topologies, with new governing constraints, order parameters, and emergent properties.

Mutually constrained and constraining dynamics can be found in "small world" networks, vehicular traffic, crowds, economies (Arthur 2020), ecosystems, and pedestrian dynamics (Gershenson 2020); they structure the universe along a different set of joints than does folk terminology. We have seen how the propagation of vehicular congestion shockwaves and the spread of infectious diseases share analogous dynamics that can be studied with simple contagion models (Saberi et al. 2020).

Synchronized and coordination dynamics display new properties and new powers. Marching in synchrony can collapse a bridge, which no individual soldier's footfalls could. It is not even the aggregate number of footfalls that brings about the collapse; it is their synchronization, an emergent property of coherence making. Context-dependent constraints thus enable new forms of order that are not merely epiphenomenal aggregates; they can "change the go of things," in Peirce's words. Herd immunity or the lack thereof is a case in point.

* * *

To recapitulate: Context-dependent enabling constraints irreversibly precipitate interactions that leave a mark. The coherence of the emergent interdependencies is that mark. Context-dependent constraints therefore drive Collier's third prerequisite of self-organization, "unity relations." In open, far from equilibrium conditions, where, as Collier (2003a) notes, there are (1) exchanges of matter and energy with the environment, (2) a source of energy, and (3) a gradient, superimposing context-dependent constraints against a far from equilibrium background established by context-independent constraints drives conditions past an instability threshold. Irreversible phase transitions produce a new coordination dynamic (Nicolis and Prigogine 1977; Kelso 1995; Turvey 1990), realized as metastable, long-range, correlated patterns of energy flow (Bejan and Lorente 2004, 2008) away from equilibrium. The constraint architecture that constitutes these patterns can span spatial as well as temporal scales and manifest as synchronic coherence and diachronic persistence (Pascal and Pross 2015). Each of these embodies spatially and temporally organized constraint regimes. Economic systems, cultures, and biological lineages are spatiotemporally organized relations generated by enabling context-dependent constraints and held together by constitutive constraint regimes.

Top-Down (Governing) Context-Dependent Constraints

As just noted, the interlocking interdependencies of constraint architectures can bring about mereological effects, top down. Remarkable new powers are among the emergent properties of the new relational landscape generated by enabling constraints. Laser beams cauterize wounds, which individual photons cannot. Molecules subtend the world of chemistry, which atoms on their own do not. A vaccinated population acquires protective powers, and entrained pendulums generate waveforms that regulate individual swings. Liquids with zero viscosity (superfluid) can flow without loss of kinetic energy; electrical resistance vanishes in superconductors.

Top-down causality is possible because *constitutive constraint architectures* of complex interdependencies generated by enabling constraints do double duty as *governing constraints*. Top down, from the invisible hand's coordination dynamics to its individual traders and firms, governing constraints in general exert control on their components and behavior in cascades of mutual constraint satisfaction, often as negative feedback loops. These cascades are typical of complex systems. They propagate information throughout the entire system as the effects of the nonlinear interactions that generated the coordination. That is how constraints tie together individual and population levels dynamics, parts, and wholes. In each case, governing constraints stabilize the possibility space within which the behavior of individual organelles, photon streams, or members of society must remain for the constrained global pattern to persist. While those governing constraints hold, the dynamic persists as an integral unit at the collective, systemwide level (despite significant changes at the component level). Unity of type endures. If overall economic conditions change (that is, if overall constitutive constraints of the economic system—manifesting as price points, for example—change), individual buyers and sellers adjust accordingly and the overall interdependencies remain steady. The coordination displayed as an overall harmonic pattern of synchronized pendulums persists in response to the top-down constraints the virtual governor applies to each pendulum's amplitude and phase. As always, such correlated and constrained patterns of energy flow simultaneously satisfy and retard the second law.

In other words, governing constraints of context-dependent coherent dynamics generated by enabling constraints keep mutually dependent relations coherent. They regulate component processes top down such that the overarching dynamic remains metastable. Collective properties of constitutive constraint regimes do so by raising or lowering barriers to energy flow, adjusting timing and activation strength, as conditions warrant. These adjustments restrict the behavior of the system and of its components or stages to a subregion of the system's overall possibility space (Salthe 1985, 2012; also see chapter 15). The emergent properties embodied in that constrained space are thereby preserved: synchrony, phase locking, coordination, and so forth hold and persist. In short, behavior controlled top down ensures it stays true to type as defined by the constitutive constraint regime that embodies its emergent properties.

Complex systems therefore regulate and control their constituents top down—as governing constraints. Like virtual governors of laser beams and pendulum clocks, governing constraints as envisioned here are distributed mechanisms of control. They are embodied in the order parameters of the interdependencies and constitutive constraint regimes established by enabling constraints. As a result, actions that issue from and are governed by those constitutive constraints remain within its more confined region of state space.

Arising concomitantly with the closure of enabling (bottom-up) context-dependent constraints that integrate separate parts into coherent wholes, top-down governing constraints (of those emergent wholes on their components) are therefore *specificatory*; they go from whole to parts, from general to particular, from constitutive constraint regime of the emergent interactional type to its actualized tokens. In the marching example, the synchronization pattern itself not only can collapse the bridge; it also acts as a global governing constraint on the tempo of the individual soldiers' cadence, whose steps become entrained into the beat. That is, the rhythm of the stepping is confined to that state space that constitutes the beat. Consequences in the real world show that these mereological influences are not epiphenomenal. Coherence as such has causal influence, just not as efficient cause.

Regulatory genes bring about change in an analogous manner. Individual transcription factors regulate by governing gene expression, in context, such that realized phenotypes preserve unity of type, even while allowing local and contextually specified variations. If governing constraints weaken, or adaptive space runs out, type can disintegrate. That is, the interlocking constraints that confer type identity will degrade, its coherent interdependencies will dissipate, and unity of type will be lost.

* * *

Reinterpreted in terms of constraint, top-down causation is real. Synergetics founder Hermann Haken (1996; Haken and Wunderlin 1988) describes lasers in unfortunate terminology as "slaving" their constituent photon streams. That is, coherent beams constrain component photon streams top down (from synchronized whole to components) such that the beam's overall alignment persists and remains diachronically coherent. Haken recently extended this idea to cities (Haken and Portugali 2021), which are complex systems par excellence. The coherent dynamics of cities arise from the way things (buildings, roads, rivers), energy, agents, and information are interlaced by constraints, including historical ones. As these become interdependent the city acquires a unique profile, an identity that embodies and enacts its history and context. During steady-state periods, the interplay among those factors is governed by a few order parameters that constrain the behavior of its residents top down: they behave like New Yorkers! Those dynamics themselves can adapt and evolve, depending on context, however. In response to a crisis, cities either undergo dramatic phase changes and reconfigure into entirely novel constraint regimes. Or they collapse. These are the transformations that make the history books.

It is important to underscore that constitutive and governing constraints should not be reified; they are not spatiotemporally other than the components and constraints from which they were formed and which they control. Because they operate from whole to parts (from extant, coherent interdependencies to token components, or to next step), governing constraints are in fact *second-order context-dependent constraints* (Juarrero 1999) that bring about specific effects and actions. They select and filter internal and external signals in light of their compatibility with existing governing constraints and accordingly restrict next steps from all possible realizations. (The anthropomorphic terms *selection* and *filtering* must be translated into the language of conditional probability: the likelihood of a particular event occurring given the requirements of multiple constraint satisfaction that describe the particular coherent dynamic in which it is entrained.)

Notably, governing constraints regulate, modulate, and otherwise control a coherent dynamic's synchronization, coordination, and so on. They control behavior such that the boundaries of the dynamic's possibility space are maintained; governing constraints also bias the likelihood of future events. In response to changing probability distributions, which in turn represent continuously adjusting constraints, constitutive constraints adaptively tweak their own profile—contextually, in response to more local and timely constraints. They adjust component processes and behavior accordingly. But they do not engage in energy transfer directly.

This is most in evidence in physiology. As noted in an earlier chapter, vessels of the circulatory system do not impart energy to the flow of blood or lymphatic fluid (Moreno and Mossio 2015; Montevil and Mossio 2015). Fluids that would ordinarily dissipate are instead confined to tubular flows along directed pathways. Blood and lymphatic vessels direct and literally canalize flow without adding energy to the processes as the heart does; vascular structure serves instead as top-down governing constraint. Should these second-order constraints (from virtual governor to components) relax, fluid flow coordination would disintegrate, and the circulatory system would dissipate on its way to thermodynamic equilibrium and death.

In individual organisms, governing constraints crosscut metabolic, endocrine, neurological, biophysical, biochemical, and other functions. They secure the emergent property homeostasis (also at many scales) by continuously adjusting its values such that constraints that govern that overarching metastable function are satisfied. Homeostasis, in short, is a dynamic version of the body's vasculature: an encompassing set of constitutive constraints that stabilize, regulate, and modulate—govern—local processes, top down. Even in the absence of tissue damage, conditions of extreme cold, starvation, or predation will facilitate responses that protect against possible loss of homeostatic function. As the organism anticipates tissue damage, homeostatic regulation activates behavior like shivering such that, overall, the organism remains within its range of viability. I submit that human cognitive and value frameworks are likewise evolved versions of top-down constraints; they, too, are second-order constraints that regulate and control behavior in terms of emergent properties.

This is possible at all only because the interdependencies formed by enabling constraints are robust and persist. Parameters set by the interdependencies govern behavior top down to ensure that persistence. Order parameters can be multiply satisfied—"appropriate behavior" can be either fight or flight, depending on the agent's history and in combination with the momentary context. In the case of homeostasis, its governing constraints damp or amplify local perturbations, oscillations, and variations as necessary, such that individual biochemical, metabolic, neuromuscular, and so forth processes covary and remain within the range of homeostasis's continuous function, as set by current local constraints as well as inherited ones. Because order parameters are multiply satisfiable, organisms can then act appropriately, fight or flee, depending on their history as interpreted at that moment, under those conditions. Later on in evolution, governing constraints embodying values, ethics, and morals and acting as second-order constraints likewise regulate and produce actions that enact those axiological constraints.

To repeat: virtual governors of synchronized pendulums and metronomes, of metabolic homeostasis, and of ecosystem and social dynamics are therefore second-order context-dependent constraints that in a distributed manner regulate and modulate energy flow such that the overall coordination dynamic—its unity of type—is preserved. That is the manner of causality of top-down influence. Standing causes in terms of long-range correlations become possible if we think of top-down causality as a second-order context-dependent constraint.

Readers familiar with complex attractors will recognize second-order context-dependent constraints. The Lorenz butterfly attractor constrains innumerable, nonrepeating trajectories; nevertheless, each run remains within the attractor's basin despite the numerous and distinct trajectories. Likewise, biological species carve out progressively individuated regions in the genus's multiply realizable possibility space, to which each variant specimen of that species nevertheless conforms (Agosta and Brooks 2020). And even very different persons are somehow clearly recognizable as Americans when they travel abroad.

Whether enacting homeostasis, a newly learned skill, or values, constitutive constraints shape new phase spaces and set the order parameters of emergent coordination dynamics. Then, acting as governing constraints, constitutive constraints operate as limiting constraints; they confine options to a subregion in state space. Like the narrowing chute of attractors in epigenetic landscapes, governing constraints carve stable basins of attraction to which the range of potential tokens and their behaviors (the actual clocks and phenotypes) are restricted. That is what virtual governors do. Governing constraints bring about these second-order, top-down effects despite being themselves products of bottom-up (enabling) parts-to-whole constraints. And to repeat, governing constraints are not other than the constitutive constraints of the complex system they govern.

Coordination Dynamics Satisfy Second Law

Constrained interdependence is energy efficient while simultaneously satisfying the second law. Interdependencies generated by irreversible phase transitions to a new form of order are paid for by an overall entropy burst. Locally, however, the new coordination dynamic has a lowered rate of internal entropy production (negentropy). As an example, the increased metastability of covalent bonds is underpinned by a net decrease in the molecule's kinetic energy. Molecular information is stored in the molecule's coherent coordination structure and interdependent constraints that subtend that metastability.

As another example of energy efficiency, infant brains, which represent about 13 percent of their body weight, burn a whopping 60 percent or so of the calories the child consumes. During an infant's first year of life, a combination of constrained development and experience trims and sculpts initially disorganized neural connections into networked pathways which are more energy efficient than the initial connections of the newborn. As a consequence of organization, the average adult's brain consumes about 20 percent of energy of the infant's usage. By streamlining patterns of energy flow, neural development induces ordered interconnections that transform complication into energy-efficient complexity.

This is a general feature of complex systems, biotic and prebiotic. Whether we consider living organelles, nonliving chemical elements, or soldiers marching to the beat of a drum, entities on their own ordinarily have more alternatives than they do once entrained into a systemwide attractor (a cell, an enzyme, or a marching army). Following synchronization or entrainment, individual trajectories are confined by the second-order relational constraints of a more energy-efficient coherent structure or dynamic, of which they are now components.

"Trimming and shaping" during neurological development is therefore not a process of random removal. It is a directed process of selection that proceeds in response to inherited and experientially acquired constraints and in reference to the systemwide coherence which it generates and preserves. The direction of fit is systematic and governed by constraints, top down; it issues from constitutive and governing constraints to appropriate action (because it is so controlled).

The approach presented in this book views top-down mereological causality as emerging concomitantly with coherence-making and hierarchization in response to constraints. Because top-down constraint is exercised in satisfaction of the emergent properties of coherent wholes, coherence-making also shifts decision-making and action control upward, to the second-order governing constraints of the interdependencies it generates. Consequently, explaining a phenomenon by reference to its fundamental particles becomes meaningless. When the *explanandum* is created by and embedded in contextual constraints, the "arrow of explanation" points upward to the systemic whole as well as down to its components (Wimsatt 1974).

* * *

Governing or top-down second order constraints were described in the previous pages as restrictive and limiting, but the last few paragraphs suggested another side to governing constraints, more accurate adjectives for which would be constitutive and stabilizing. Because coherence making—the formation of a new type of entity—underlies dynamic metastability, top-down governing constraints are not a priori negative. Without the stabilization they provide, exponential growth precipitated by enabling constraints would spin out of control like the positive loop gain that makes electronic speakers screech.[8] Moreover, governing constraints hold together constitutive constraints over time and thereby allow the interdependencies in question to persist as themselves far longer than otherwise. Without their stabilizing ratchet, evolution would be impossible. Interdependencies would decohere and the second law would quickly disintegrate any incipient order. Governing constraints both stabilize and preserve achieved coherence ready for the next step.

As an example from social science, limitations and prohibitions enumerated and described in a nation-state's founding documents establish governing constraints on its member states and citizens ("Powers not delegated to the United States by the Constitution, nor prohibited by it to the States are reserved to the States respectively, or to the people."). These constraints delineate a space of possibilities (Artigiani 1994) that limit what individual persons, or states, can or cannot legally do. These settings increase the odds that residents, citizens, and states remain aligned overall as members of the body politic.

Belonging to a body politic also opens dimensions that afford different and qualitatively novel roles and opportunities for its members. Being recognized by the "community of nations" does the same for a nascent state.

New rights and opportunities appear concomitantly with new responsibilities and constraints. Without the emergent affordances offered by legal and judicial platforms through their enabling and governing constraints, individuals cannot be spouses, mayors, or conscientious objectors, for example. Nongovernmental organizations cannot form. Individuals cannot participate in a banking system. As Locke, Hobbes, and Rousseau well understood, such roles (and authorizing powers) are unavailable to a motley group of people prior to coalescing as a social organization of a particular sort (organization being defined in terms of those interlocking constraints that establish and stabilize its coherence).

Similarly, genetic material captured in an amino acid, amino acids captured in a protein, and mitochondria captured in cytoplasm have fewer degrees of freedom than as isolated biochemicals. Amino acids are constrained components of the coordination dynamic that constitutes a protein. The wider set of interdependencies (which we call proteins), however, acquires properties and functionalities the amino acids individually lack. Molecules have properties and powers that atoms singly do not. Full-fledged organisms can access more dimensions and degrees of freedom than isolated RNA molecules can. Enabling constraints generalize; the new degrees of freedom these interlocking interdependencies afford are realized as emergent properties and powers. Traits and powers are not new things; they are *condensates* (Burgauer 2022a) or *foci* of constrained relations endogenously generated bottom-up. In turn, and acting as second-order governing constraints, they stabilize and specify but neither determine nor freeze behavior by restricting the tokens' degrees of freedom to within that newly organized constraint regime.

Emphasizing this point reiterates the fact that authoritarian and arbitrary fiats of dictators, field marshals, and some CEOs are not the kind of top-down or governing constraints envisioned here,[9] which are naturally formed distributed and second-order constraints that arise from first-order, contextual constraints on individual clocks, photon beams, biochemical reactions, and living things. Synchronized oscillating patterns of pendulum clocks on a shelf, coherent laser beams, trophic ecological flow patterns, circadian rhythms, the vascular system, and more recently mores and norms enacted in social practices and traditions are what we have in mind when we refer to top-down governing constraints.

The next few chapters focus on catalysts, autocatalysts, feedback loops, and a unique type of constraint, *closure of constraint*. This folding-back-on-itself characteristic of recursive and iterated loops of processes and constraints, of which autocatalytic cycles are a prime example, is central to the generation of coherent dynamics. It merits a separate chapter.

7

Catalysts, Loops, and Closure

Catalysts and feedback loops play as important a role in economics (Arthur 2020) and finance (Mazzucato 2020) as do biochemical catalysts and metabolic loops in biology.

Although the damping and stabilizing power of negative feedback loops has long been understood, positive feedback like the disagreeable screech of unchecked positive feedback in audio systems was taken to be exclusively problematic. Discovering how feedforward, positive feedback, iteration, and recursion, acting as constraints, can precipitate phase transitions to coordination dynamics with emergent properties has been among the important contributions of complexity theory. A key point in our story, once again, is that context-dependent constraints weave together interlocking interdependencies without directly injecting energy. The self-reinforcing and coherence-making powers of autocatalysis and feedback loops are central to extending the received understanding of causality with the notion of constraint.

Catalysts and Feedback Loops

Catalysts speed up chemical reactions by lowering barriers to energy flow and thereby facilitating irreversible interactions without being consumed themselves. They lock in new interdependencies and create new constraint regimes. Catalysts enable coherent sequences and dynamics to emerge. Metaphorically, *catalyst* also describes persons or processes that clear the way for major transformations.

As always, catalysts are pertinent to our story because they function as context-dependent constraints. They do not undergo permanent alteration; they are conserved. Enzymes, which are catalysts, are responsible for most biological reactions, but neither as reactants nor products. Although most enzymes are proteins, some RNA molecules can function as enzymes as well. They too bring about effects as catalysts; they make available paths along which energy can flow; doing so facilitates other processes. In the presence of catalysts, certain reactions, in a certain direction, become more likely, happen faster, sooner, or more efficiently. As a result of these contextually constrained changes in the likelihood of their occurrence, flows of energy (or vehicles, or information) are accelerated and oriented toward certain directions and away from others. The reason super glue will bond to your fingertips if you are not careful is because trace amounts of acid on the skin function as a catalyst that quickly hardens the glue.

Folding-back-on-themselves processes such as feedforward and feedback loops are also catalysts. Iteration and recursion are two such examples. Often used as a synonym of *repetition*, suggesting a simple context-independent constraint, *iteration* feeds back information from the output of one run into the initial conditions of the next. Iteration thus acts as a temporal constraint that changes the likelihood of the next output. The process repeats.

In *recursive iteration*, full sequences are fed back on themselves. This looping causes processes and sequences to become self-referential; recursive iteration blurs the distinction between parts and wholes. As an example, software used to predict protein folding feeds output back into the system (Jumper et al. 2021). As a result, the overall dynamic itself becomes a template, a catalyst that relies on context to guide folding.

In other words, iteration and recursion feed information from the context back into the next sequence as newly initialized conditions and constraints. Such looped and contextually constraining and constrained interactions effectively import spatial and temporal information about the world into those processes and their properties. As a result, the processes become interdependent and covary with events in the world. In machine learning software, loops of backpropagation algorithms change the weights of middle-layer connections such that they incorporate important, meaningful, and real-world distinctions in their attractors and constraint regimes. Over time, the system's input and output become increasingly tuned and calibrated to its context and its moment. Qualitatively new results emerge from such context dependence (Hinton and Shallice 1991; Juarrero 1999).

Attractors in possibility space have a vectorial quality. By extruding into the context, feedforward processes modify themselves in anticipation of expected conditions; the loops thus make the attractors anticipatory. Feedback and feedforward loops of energy flows (Rosen 1985; Hoffmeyer 2017) that generate attractors as depicted in figure 4.1 establish pattern completion tendencies in dissipative structures, biotic and prebiotic alike. As we saw in the case of inflammation, they are even able to accommodate anticipated threats to the regulatory function of homeostasis. Terms like *modify themselves* and *anticipate* should be understood as changes in the probability distribution of events in possibility space.

Prairie grasslands offer an ecological case study of this sort of anticipatory cooptation (Allen and Starr 1982; Gare 2019). Grasses compete with both flowering plants and grazing animals such as horses; competing with the former is the more energetically expensive of the two. By putting out *apical meristems*, a growth region at the tip of the roots and shoots, grasses set out an enabling constraint that makes horses more likely to graze. As a virtuous cycle is established, horses become integrated into a now more comprehensive constraint regime. Previously predators, horses are now folded into a more encompassing context, a new niche or habitat with new constitutive and governing constraints.

It is important to note that recursion and iteration are possible only after temporal dependencies (straightforward sequences) have already formed in response to enabling, context-dependent constraints. That is, recursion and iteration are not possible without previously constrained ordinal relations. That said, however, when the last step of a sequence feeds back to become the first in the next iteration, the looping creates self-referential configurations and nonlinearities. Nonlinearities generate multiscale and multidimensional interdependencies.

Iteration has these kinds of significant effects in olfaction, for example; emotions experienced when first exposed to a particular odor are fed back as initial conditions when that subject is next exposed to that scent. The earlier correlation acts as a constraint on the present sensory input and the odorant is experienced as a fragrance or a stench depending on the valence imprinted the first time (Freeman 1991a, 1991b; Barwich 2020). Over time, subsequent experiences with that odorant can feed back into the established dynamic and modify it accordingly, but the interdependence born of integrating different sense modalities into coherent units can persist, with modification, throughout a subject's life. This contextual integration does not make the perception of a particular odor subjective and arbitrary; it makes it a real but relational, contextual property of a higher-level coherent dynamic (Barwich 2020) coded as a combination of two constraints, temporality and valence.

Iteration and recursion are therefore unlike simple recurrence, repetition, and replication described in the next chapter. Iteration and recursion are hybrid constraints. Both take systems farther from equilibrium and therefore qualify as context-independent constraints. But by feeding real world information back into the process, iteration and recursion also weave context, history, and the subject's own actions into a more encompassing coordination dynamic—the spatiotemporally more extended interdependencies of a new context. In this role, iteration and recursion function as context-dependent constraints. Once recursive or iterative loops close thanks to integration by enabling constraints, real-world spatiotemporal information becomes embodied in a qualitatively distinct set of interlocking relations with novel properties.

In short: ever more complex and extended interdependencies form when ordinal relations fold back on themselves, when the output of a temporally coded sequence conditions the first step of the next iteration. By folding processes back on themselves, recursion and iteration generate multidimensional interdependencies bound together into multiscale yet coherent wholes. It is these interdependencies that underpin covarying relations between parts and wholes. Temporal constraints of lineage and ontogenetic development intertwine in this manner. They illustrate how recursion, a feature that bedeviled modern science and philosophy, returns as a central driver of increased complexity and a source of emergent dynamics, properties, and powers. The next section delves deeper into *autocatalysis* and feedback loops as enabling context-dependent constraints.

Closure of Processes (Catalytic Closure)

Cycles of energy-driven processes, such as the hydrologic (water) cycle, are well known. We focus now on cycles that implicate constraints more directly. The citric acid (Krebs) cycle is an eight-step sequence that involves several enzymes (catalysts) in a series of chemical reactions.[1] Iteration turns the sequence autocatalytic when the product of the last reaction catalyzes the first. By closing the loop, the cycle repeats and persists. Autocatalysis and feedback loops are therefore self-reinforcing mechanisms; they self-organize and self-maintain. Water, adenosine triphosphate (ATP), and Krebs cycles are instances of process or reaction closure. Moreno and Mossio (2016) note that Stuart Kauffman calls the last two cases examples of catalytic closure. The remainder of the chapter focuses on autocatalysis and feedback loops as enabling context-dependent constraints.

Autocatalytic Closure

The Belousov–Zhabotinsky (BZ) chemical reaction takes single step autocatalysis one step further. Disregarding the actual chemicals involved, the BZ reaction can be schematized as follows (Jantsch 1980):

$$A \rightarrow X$$
$$B + X \rightarrow Y + Q$$
$$X \rightarrow P$$
$$Y + 2X \rightarrow 3X$$

In this schema, chemical X catalyzes and is catalyzed by different reactions. The fourth step, however, is noteworthy: the product of the reaction, X, is also necessary to realize the reaction itself. This makes the fourth step autocatalytic. This reaction reproduces the conditions necessary for catalysis to take place.

Most significantly, however, the BZ reaction realizes two intertwined autocatalytic cycles. In addition to the fourth step, the sequence "as a whole acts like a catalyst which transforms starting products into end products" (Jantsch 1980, 33). When the fourth step folds back on itself, that is, it also closes a second loop of constraints, one that folds back onto the first step in the cycle. This more encompassing autocatalytic *hypercycle* (Eigen 1971) includes the first loop (the fourth step) as one of its components. By renewing itself continuously in a recursive process of individual reactions, the hypercycle itself becomes an enabling constraint that induces its own production and maintenance. Mereology with a vengeance.

In short, new regimes of multiscale interdependencies and correlations emerge from recursively iterated autocatalytic constraints. Unlike convection cells (which depend on externally set boundary conditions), the BZ's dynamic realizes a more complex and mereological set of interlocking interdependencies by endogenously regenerating the conditions and constraints necessary to form and self-maintain. Its self-reinforcing dynamics are less dependent on externalities, earning it the labels self-generating and self-maintaining—because self-reinforcing.

Dissipative processes like the Rayleigh–Bénard cells and BZ reactions described earlier are autopoietic. Chilean biologists Humberto Maturana and Francisco Varela (1979) coined the term *autopoiesis* (self-organization) to describe the self-production of cells, a sort of "collective autocatalysis plus spatial individuation"[2] (McMullin 2000). Autopoiesis does not replicate as such. Instead, because the constraints are context dependent, it persists and individuates.

When sequences of reactions (including autocatalytic steps) become looped in this manner, the overall dynamic accelerates exponentially. Like other dissipative structures described earlier, exponential growth can drive the dynamic over a threshold of instability, precipitating a transition to a new phase characterized by long-range and metastable correlations that are visible in novel properties. In Bénard cells, these appear as rotating columns of fluid; in the BZ reaction, they manifest as the signature population-wide properties of colorful, spreading waveforms. The details of each convection cell and BZ reaction are distinct. Because they are path-dependent, the exact pattern differs each time depending on minuscule differences in initial conditions as well as fluctuations and perturbations encountered along the way. Contextually constrained path dependence underpins variation, but despite such multiple realizability, autocatalytic closure ensures that the macroscopic flow pattern (the rotating column or wave) persists overall by simultaneously self-generating and self-maintaining multiscale loops of self-reinforcing constraints on constituent processes. They self-stabilize as a result.

The BZ chemical reaction is significant because, more so than physical convection cells, hyperloop-creating constraints revive the specter of self-cause, that bête noire of Western philosophy and science. The two autocatalytic cycles in the BZ reaction intertwine to self-produce a set of mereological constraints. The hypercycle as a whole stabilizes and governs its individual reactions (including the locally autocatalytic fourth step) which, when combined, enable the hypercyle in the first place. This recursive and Escher-like "strange loop" (Hofstadter 1979) makes BZ hypercycles exemplars of Kant's self-organizing self-cause—"a form of causality unknown to us" (Juarrero-Roque 1985).[3] BZ reactions illustrate how emergent and coherent wholes are dependent on the parts that generate them as well as how the behavior of those parts becomes dependent on the whole in which they are entrained. This integrated regime of mutually dependent hypercycles brings into existence a coherent dynamic and the conditions that make it possible; critically, this self-reinforcement also shields that coherence from quick dissipation—by keeping it away from thermalization. Both are essential for its persistence in time. The hypercycle persists as itself because its conserved constraint regime and component alignment retard the second law through organization.

* * *

In their various works, Montevil, Ruiz-Mirazo, Moreno, and Mossio (Mossio 2013; Ruiz-Mirazo and Moreno 2004; Montevil and Mossio 2015; Moreno and Mossio 2016) maintain that self-organization and autopoiesis in physical and chemical dissipative structures still depend on the externally set (context-independent) boundary conditions over which they have minimal influence. The constraint regimes they realize, that is, are not yet self-constraining. As such, they are only precursors of full autonomy and self-determination. For those two novel properties to appear, closure of constraints is necessary (Montevil and Mossio 2015; Moreno and Mossio 2015).

Constraint Closure

In the several papers and books referenced earlier, Montevil, Moreno, and Mossio argue that a different sort of closure—closure of constraints—supports a measure of agency that renders living things self-determining and autonomous. They are autonomous because, by realizing endogenously self-determining and persistent loops of constraints, not simply loops of processes, they not only self-assemble and self-maintain in response to externally set constraints. They self-constrain. It is in virtue of interdependencies bound together as loops of constraints, not just of processes, that living things also realize a novel form of constraint, constraint closure. The three authors claim that *closure of constraint* generates a novel type of phenomenon, *Life* (Mossio 2013; Montevil and Mossio 2015; Moreno and Mossio 2015).

Closure of constraint binds together several local and shorter catalyzed reactions into an extended and catalyzing loop, all without injecting additional energy into the process. The loop itself becomes a virtual governor, a second-order constraint that catalyzes other more local constraints (not just other reactions) and thereby generates an extended and mutually interdependent dynamic, a hypercatalyst (not just a hypercycle). This overarching dynamic becomes a self-reinforcing governing and catalyzing constraint. In doing so, constraint closure turns the process self-determining.

Self-determination as self-constraint is absent in systems characterized solely by process or catalytic closure[4] for which externally set boundary conditions are still significant. As articulated by Montevil, Mossio, and Moreno, the word *closure* in closure of constraint refers to a specific mode of dependence between constraints whereby recursion in the chain [of constraints] "folds up and establishes *mutual* dependence" (Moreno and Mossio 2015, 20) among constraints. To wit:

> Formally, "a set of constraints C realizes closure if, for each constraint C_i belonging to C:
>
> C_i depends directly on at least one other constraint of C (C_i is dependent);
>
> There is at least one other constraint C_j belonging to C which depends on C_i (C_i is enabling)." (Moreno and Mossio 2015, 20)

"Closure refers to an organisation in which each constraint is involved in at least two different dependence relationships in which it plays a role of enabling and dependent constraint, respectively. The network of all constraints, which fit the two requirements, is . . . collectively able to self-determine (or, more specifically, self-maintain . . . through self-constraint)" (Moreno and Mossio 2015, 21).

Visually tracking the simultaneously constraining and constrained processes as illustrated in figure 7.1 lays bare their mereological interdependencies. As can be appreciated, constraints can operate at different time scales. Constraints C_{1-5} facilitate reactions on a brief time interval. In contrast, the overarching loop of constraints that cycles through reactions involving $C_2 - C_4$ spans (and is conserved throughout) a longer interval, τ_2 through τ_4. It realizes a spatiotemporally extended, cyclically repeating set of interdependencies. As a self-generated and self-generating hypercatalyst, the cycle functions as a novel second-order constraint that persists despite turnover in its components. Let us run through the steps.

In this schematic illustration, the sole independent constraint, C_5, is an enabling constraint that shows that biological systems are open systems (Mossio, email communication 2021). Constraints C_{1-4} are dependent constraints. As an individual catalyst, C_4 directly depends on C_2, on which C_1, a by-product, also depends. Catalyst C_3 depends on C_4 and C_5.

Each of these constraints, however, is also a generative or enabling constraint: catalyst C_4 generates C_3, which generates C_2, which in turn generates C_4. Unlike the BZ reaction, no individual reaction in figure 7.1 is autocatalytic: no individual catalyst catalyzes a reaction that produces that catalyst. Collective autocatalysis is realized in the overarching cycle of constraints C_{2-4} (outlined as a black oval). The illustration shows how the cycle as a whole constitutes a coherent unit, simultaneously a dependent and an enabling constraint. Its formation depends on a specific sequence of individual catalysts, C_{2-4}. Some are enabling; some are governing, but it takes the closure of the encompassing loop of constraints (when constraint C_4 directly catalyzes C_3) to regenerate the sequence of reactions on which the loop's very formation depends. By generating and regenerating the constraint interdependencies that persist and delay dissipation, such recursive and multidimensional organizations display qualitatively different properties than physical and chemical convection cells. Processes that realize self-constraint are self-determining. They represent a major transition in evolution (Maynard Smith and Szathmary 1995) brought about by recursively organized constraints.

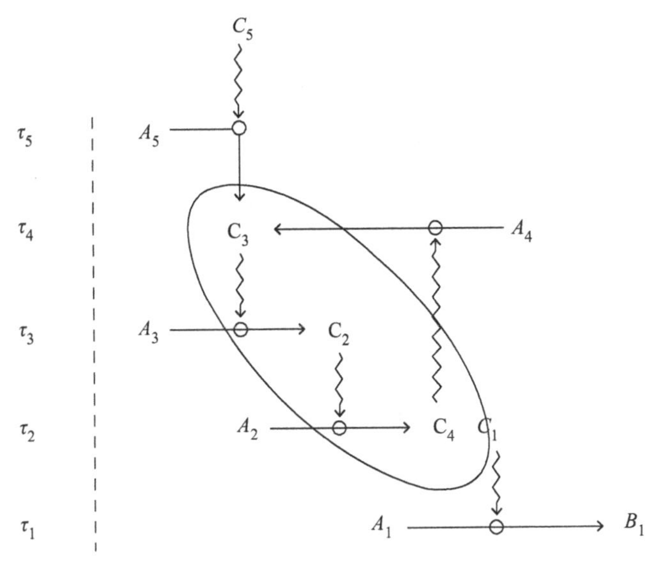

Figure 7.1
Schema of constraint closure. *Source:* Montevil and Mossio 2015, 186

The received view of causality contends that wholes are nothing but aggregates and therefore epiphenomenal and that efficient causes are other than their effects. We cannot overemphasize the point that interdependencies that realize closure of constraint are not spatiotemporally "other than" the local constraints, and yet as a coherent unit, the loop exerts constraining influences on its component reactions that individual constraints do not. The effect of its influence is the realization of a higher-order dynamic with the novel property of self-constraint. Clearly falsifying the classical prohibition against mereological self-cause, interdependence among constraints reveals a relation that is simultaneously constraining and constrained, from whole to parts and parts to whole.

A crucial point stands out with respect to autopoiesis, autonomy, self-constraint, and self-determination as just characterized. As explained in the first chapter, the classical notion of independence centered on primary properties; it refused any role to interactions and relations with either the environment or history. In contrast, context-dependent constraints in open systems far from equilibrium underpin self-generating and self-maintaining interactions and relations that are nevertheless enmeshed with the environment and history.

Interdependence with spatiotemporal context is at the basis of autonomy, autopoiesis, self-constraint, and self-determination. These terms describe processes that generate and maintain emergent systemwide organizations by virtue of constrained interactions among their components and with the real world. In contrast to classical essential traits, then, contextually constrained interactional types and their tokens as envisioned here are emphatically not independent of the rest of their world or their past. To the contrary, they integrate and realize various degrees of interdependence with the world and with history. They embody and enact coordination dynamics, niches, and habitats, as well as the umwelt of living things, but they are neither determined by nor absorbed into the context.

This mereological interplay is central to discussions about identity.

Statistical and Semantic Closure

As Moreno and Mossio note, American biologist Howard Pattee characterized closure of constraints in two ways. Initially, he described it as *statistical closure* (Pattee 1973). The word statistical (in Pattee's characterization) refers to selective loss of detail with respect to the underlying dynamics. Pattee considers loss of detail a product of epistemic simplification, which becomes necessary because of the impossibility of keeping track of a complex system's myriad microdetails.

From the perspective presented in this book, in contrast, the "selective loss of detail" of epistemic classification tracks nature's ability to generate coherence through constraints, coherences that embody emergent properties at each turn. Montevil, Moreno, and Mossio's point is that Life is the property that emerges from the variety of generalization realized by constraint closure. Arnellos and Moreno (2022) meticulously track the emergence of specific functions like perception from the operation of particular constraint regimes at precise moments in time. The goal of this book has been more general, to describe the emergence and persistence of the constraint regimes themselves.

Consider our usual suspect, homeostasis. Describing a combination of metabolic, endocrine, neurological, and other processes as implementing homeostasis glosses over the microdetails of the diverse and individual biochemical reactions that realize that dynamic. From the perspective of complexity, the apparent loss of detail captured by that term is not solely an epistemic or methodological issue; nor is it due to ignorance. This book has argued that the possibility space and parametrized constraint regime that constitute the overarching dynamic we call homeostasis are real; the term simply compresses the myriad interdependencies of real and constrained interactions into a common noun that captures the salient parameters and emergent properties of those interdependencies.

When depicting contextually constrained open systems far from equilibrium, I submit, statistical classifications and the common nouns used to describe them (homeostasis, chemistry, vorticity) identify coordination dynamics generated by context-dependent constraints. Lasers, convection cells, chemical processes, homeostasis, and even meaningful human actions are not clumpings and de-clumpings; the common nouns describe salient emergent features of overarching relational interactions generated by enabling, context-dependent constraints; such interdependencies are characterized by novel relational properties and order parameters that constrain component processes and behavior.

In sum, I propose that contextually constrained interdependencies (interactional types) can be epistemically simplified and described with common nouns only because those coordination properties and order parameters they identify are real, relational, and multiply realizable interdependencies generated by contextual constraints. They possess emergent properties over and beyond the properties of their components. The nouns refer to the interlocking constraints of a range of tokens of the same type. The energetic streamlining effected by those interlocking constraints is why the mereological dynamic is also called "self-simplifying" (Pattee 1972b).

In sum: what descriptors such as homeostasis and congestion shockwave—or thoughts and feelings—lose in detail about the microlevel they gain by identifying salient and systemwide properties of contextually formed coordination dynamics. These interdependencies and emergent properties are captured in new-order parameters. The second-order constraints those interdependencies embody constitute a new type of entity: they also serve as its governing constraints.

Moreno and Mossio also note that Pattee's statistical classification—the loss of detail—refers to control constraints, here called governing constraints. These second-order constraints have been described as a system's internal map and clock (Goodwin and Cohen 1969); they calibrate, regulate, modulate, and time the activation and suppression of component processes (like pendulum clocks mentioned earlier) such that the emergent property is enacted. As a self-contained and coherent hyperconstraint formed in virtue of interlocking lower-level constraints, the closed hyperloop C_{2-4} illustrated earlier, for example, constitutes a real governing (second-order, context-dependent) map and clock for the individual components and processes that comprise it, and for the actions they perform. As such, emergent properties can be said to govern the behavior and actions of complex systems, top down. In analogous fashion, molecular dynamics set the map and clock of chemistry; cultural mores and traditions set the map and clock of social practices; and initialization sets the map and clock of IT platform interactions.

Acting as hyperconstraints, in other words, loops of constraint establish new spatiotemporal contexts within which component reactions are now embedded and which give them meaning. The context is, in effect, a new possibility space with a new map and clock. It is also a new hierarchical level of dynamical organization with novel properties governed by a distinct constraint regime. Statistical classification is therefore not only a "linguistic description" (Pattee 1978); it is a compressed characterization of the governing constraints that hold interdependencies together—their self-simplification into a coherent dynamic. And to reiterate, despite its novel properties and powers, the embedding context (the hyperloop in this case) is real but not "other than" the reactions that make it up. By calibrating and controlling the loop's components, conditions, and timing, constraint closure keeps the multiply realizable interdependencies within a tolerated range of stability and viability—it preserves unity of type. The real spatiotemporal context it establishes disambiguates and specifies those circumscribed possibilities.

* * *

A decade or so after his first comment on statistical closure, Pattee (1982) proposed a different, semantic interpretation of what is now being called closure of constraints: *semantic closure*. Using the example of transfer RNA (tRNA), he noted that "semantic closure arises from the necessity that the translation molecules [tRNA] are themselves referents of gene strings" (Pattee 1982, 333, quoted in Moreno and Mossio 2016, 21). Biosemiotics proposes that strings of genes acquire a symbolic role and become meaningful when tRNA "recognizes" and realizes the gene's meaning (its reference) by synthesizing the appropriate protein. The label "appropriate" implies that the same functional protein is degenerate: it can be synthesized from a variety of amino acid strings, depending on context, including timing. Mutatis mutandis, the same could be said for the context-dependence emphasized in the theory called virtue ethics (MacIntyre 1981): actions are morally appropriate only in a given context. The same behavior physically described might not be ethical, in a different context.

Messenger RNA (mRNA)–based COVID-19 vaccines do not contain antigens; mRNA acts as instructions for constructing those antigens, *under the right conditions*. This last phrase is important: the most difficult part of developing the mRNA COVID vaccine was ensuring its protection and successful delivery to cells. Accomplishing this required adding a buffer that would protect the instructions until the right conditions are encountered. Timing and placement as set by the organism's various maps and clocks change properties and powers. So whence the capacity to recognize "as appropriate under these conditions"? Complex systems have been described as energetically open but organizationally and informationally closed.[5] How can Pattee's interpretations of syntactical and semantic closure be simultaneously implemented?

Several chapters in this book have already proposed that coherence knit by context-dependent constraints enacts and embodies real features of their world. Multiscale feedback processes import real and meaningful features of the world into the system's constraint architecture; they do so by importing the "statistics of their world" (Humphries 2021) into the dynamics of the interdependencies they generate. As a result, heritable constraints first set the unity of type that delimits a given possibility space (Brooks and Wiley 1988, H_{max}). These constraints determine features that are conserved despite perturbations. Then, local conditions add context-dependent constraints that circumscribe the range of currently viable tokens. Potential diversity becomes progressively specified and realized at each decision point. Those conditional probabilities import the real meaning to be recognized and synthesized at the right moment—for example, the appropriately functional protein, or the morally correct action. Genes are material embodiments of the constrained syntax and semantics of life. Moral norms and values are the symbolic embodiments of the constrained syntax and semantics of social organization (Artigiani 2021).

Interwoven context-dependent constraints, in other words, fold in real-world semantics into the very syntax (the constraint regime) governing the realization and path dependence of actual tokens (Juarrero 1999). Syntax and semantics, that is, are two aspects of a common process of integration and selection that are separable only at the population level. Chapter 15 describes neuroscience experiments that show how this integration is produced in the brain.

Semantic Attractors

Dyslexia provides an interesting case study for how syntax realizes semantics by incorporating the statistics of the world into a dynamic network. Over thirty years ago, cognitive scientists (Plaut and Shallice 1991; Hinton and Shallice 1991) trained artificial feedforward neural networks to read words using back-propagation algorithms. These algorithms use iteration and recursion to reduce error by updating values of the initial conditions.

When the scientists lesioned the networks above the feedback loops, the networks' output (its printout) committed surface dyslexia-type errors, simple errors of transposition. Humans with surface dyslexia commonly transpose letters—by reading or writing *bed* for *bad*, for example. Meaning is not implicated in such errors. However, when lesioned below the feedback units, the artificial neural networks made errors like those of human subjects with deep dyslexia: the printout spelled out words whose letters differed significantly from the letters which the software was presented, but the words preserved the input's meaning. For example, when lesioned below the feedback loops, a trial run presented with the letters *bed* might print out *cot*; when shown the letters *band*, its output might be *orchestra*. That is, the weights and connections of the neural network's middle layer preserved the meaning of the words in the input.

When asked to explain this astonishing output by a silicon-based system, the authors hypothesized, "The network must have created a semantic attractor." Attractors, as has been explained, are constrained dynamics that represent a set of values toward which the system tends to return. In this case, that set of values embodies and enacts semantics.

On this account, then, meaning is therefore not a picture in a homunculus's mind; nor is it an uninterpreted lookup table. Just as a snowflake's structure embodies the real-world conditions under which it was created, the weights and connections of a neural network's dynamics established by contextual constraints are locked in to the regularities of its training set's constraints. In this case, those regularities embodied real world meaning. The constraints of the neural network's attractors therefore printed out words that conserved those real world regularities and preserved meaning.

Semantic attractors are not things; they do not control top down for meaning through energetic transfer. They initialize and reset starting conditions and conditional probabilities that emerge and covary in response to ongoing real world constraints, those of the training data set and backprop algorithms. The conditional probabilities of these combined constraints weight the middle layer such that the likelihood of certain outputs increases, and others decrease. These conditional probabilities can be represented as hillocks and troughs in constraint space. Individual trajectories within the newly organized (semantic) constraint regime enact those conditional probabilities. In the example presented, these weights and connectivity capture the input's real-world meaning. Although multiply realizable, each actual run remains within the dynamic's *basin of attraction*—the type's constraint regime. If the network is lesioned downstream from the feedback loops that integrate the connection weights of its middle layer, the attractor's output remains governed, top down, by the constraint regime of its overarching basin of attraction. As a result, the printout preserves the emergent properties (the type-level meaning) and order parameters of the attractor. If the network is lesioned upstream from the feedback loops, no meaning is transmitted.

Biosemiotics avoids ontology (although see Garre 2019). From the perspective of complexity theory, however, feedback loops can be viewed as real and enabling because, as parts-to-whole (or bottom-up) contextual constraints that lock in to real-world interdependencies, they ontically partition the possibility space of initially separate processes into coherent—and real—types of contextual and local dynamics, with emergent properties. As has been emphasized throughout this book, coherent constrained dynamics are organized regions of state space realized as molecules, living things, and human organizations.

Loops of constraint and recursive processes also generate new and real forms of control or governance—mereological constraints. Operating whole to parts (or top down), second-order constraints of those types control behavior such that the network's printout embodies the emergent properties of that constraint regime. Semantics, in this case, homeostatic metastability in others. In context, through cascades of negative feedback, second-order constraints control component entities and processes top down such that these embody the constraint regime's emergent properties and thereby remain invariant despite the variation of the actual token.

* * *

This chapter has shown how, in physical and chemical processes such as Bénard cells and BZ reactions, loops of constraint self-organize, self-maintain, and persist. These are precursors of the more complex dynamics of self-determination and self-constrained autonomy that will emerge with living things. The differentiated coherence of each of these complex interdependencies generated by constraints allows multiple energetic pathways to realize the same overarching set of governing constraints. Top-down regulation by those governing constraints keeps the system's behavior within its basin of attraction even as local context-dependent constraints specify the details of realized tokens (the specific features of an individual organism) and tailor those details to actual circumstances. Multiple realizability thus makes a significant difference in the evolution of self-organization, self-maintenance, and self-constraint; it produces astonishing transitions to new mechanisms of governance and control that are increasingly autonomous and self-determining.

8

An Abundance of Constraints

Coherence-making by constraints is one thing. Coherence maintenance and persistence is another. As implemented by default governing constraints, maintenance and persistence presuppose relatively steady background conditions. For example, regular solar flares and neutron star collisions light years away do not threaten the planet's atmospheric dynamics, whose governing constraints are metastable to such routine fluctuations. Earth "treats" those slower cycling rates as "constants" (Allen and Starr 1982; Salthe 1985, 2001).

But conditions inevitably change, and existing governing constraints that keep local systems in balance can become precarious, unable to maintain coherence. Interlocking constraints that realize metastability will decohere when either context-independent or existing constitutive constraints are sufficiently perturbed. Interdependencies that were once stable and effective become threatened. Eventually, all coherence is threatened by the second law and the inexorable march of time; unusually rapid and severe changes can render its governing constraints ineffective and even unresponsive to changing conditions. Any responses the system does manage to produce will likely be too slow, stale, and sluggish—mismatched to current conditions. Resulting decompensation is as true for human organizations as for natural ecosystems (Woods 2016, 2018).

Warning signs of impending disintegration appear. In natural and simulated systems alike, unusually large fluctuations and rapid oscillations presage an approaching instability. Trends in the earth's climate offer stark evidence of the threat today: extreme wildfires and record temperatures and rainfall are indicative of the planet's stressed governing constraints. Ecosystems become saturated and depleted. When existing governing constraints are no longer sufficiently coherent (or sufficiently adequate to current conditions) to preserve their interdependencies, organisms, ecosystems, and cultures alike cannot cope. The possibility of disintegration looms. Stars burn out. In living things, senescence sets in.

Changes in constraints at scales closer to those of the threatened system can directly alter its chances of survival. Supplementary constraints tacked on to default enabling and constitutive/governing constraints can reinforce existing organization and thereby support the system's persistence. They prolong its metastability in the face of strong instabilities.

Adding extra oomph to existing constraints of all stripes keeps gradients from equilibrating and interlocking interdependencies from decohering. Add-on support like new varieties of memory and regulatory constraints can keep systems sufficiently stable and prolong their existence. Others constraints like "more-making" (Eldredge 2015), buffers, and isolation can also keep systems from thermalizing by holding conditions away from equilibrium. Density, motility, and migration (Agosta and Brooks 2020) are yet other constraints that access new energy gradients with fresh resources. Some of these add-on constraints appeared only with the origin of life; others existed much earlier. Some are context independent; others are context dependent. This chapter presents a cursory survey of common add-on constraints that prolong persistence.

* * *

Repetition, replication, regularity, redundancy, and reproduction—"more-making" (Eldredge 2015)—are examples of such constraints.

The common prefix is not coincidental. Serving as context-independent constraints, various forms of again, again, and again increase the numbers of those entities or processes in the state space. By increasing their percentage in the overall population, the resulting density deforms the state space and takes systems even further from equilibrium with the environment. More, more, more increases the likelihood that those systems will not quickly dissipate in the face of perturbations and fluctuations. It also increases the likelihood that their traits do not quickly degrade.

Repetition

As a case in point, short nucleotide sequences often repeat in the genome. The gene for the flower of Gregor Mendel's peas has two repeats in its pigmentation gene: each repeat codes for only one of the flowers' two colors, white or purple. When the number of repeats varies, each repeat can represent an inherited allele. This reinforces the faithful persistence of that pure line trait. Repetition was also among the earliest human communications techniques to improve fidelity of transmission in a noisy medium. Native American smoke signals and the talking drums of the Yoruba people are long distance communication codes that date back to antiquity. By coming at evenly spaced intervals, regular (hence improbable) angular rotations of pulsars, for example, reveal the presence of two constraints, magnetic field and local density of matter. Lighthouse makers like the family of author Robert Louis Stevenson likewise understood that regular flashes of light convey information because they are regular. Repetition and regularity on their own do not impart energy; like timing, they preserve and transmit information by deviating from equiprobability and randomness. As constraints, they convey information of constraints.

In human organizations, too, regular reenactment of constrained behavior preserves social cohesion and adherence to religious and cultural teachings. Yearly rituals and their attendant prescriptions and proscriptions keep cultures and religions alive by reinforcing existing constraints. Personal habits and routines also reinforce constraints. We misplace keys less often by leaving them on the same console. Practice might not make perfect, but it tends to consolidate it by preserving constitutive constraints in nonequilibrium.

Mechanical engineers rely on a related mechanism,[1] *redundancy,* to improve system reliability. By adding extra components that can compensate for any loss or damage to the main one, redundancy aims to secure *fail-safe* conditions and keep the system from disintegrating. If one signal does not get through or a component fails, another takes over. In both information theory and engineering, then, more, more, more increases the likelihood that message transmission and functionality do not degrade. Redundant measures are, therefore, context-independent constraints that keep systems away from equilibrium, despite perturbations. They shield coherence and functionality from equilibration. Redundancy is expensive, however, with diminishing returns. If the first signal gets through or the primary component works as intended, redundant units are superfluous and energetically costly.

Replication

Replication, making a copy of something, is a variety of repetition; it is another form of more making, a context-independent constraint. Unlike redundancy, replication does not just add an extra existing copy; it makes new copies. *Mitosis*, for example, produces two identical daughter cells. By doubling the number of tokens at each turn, mitosis takes conditions even further from equilibrium.

Mitosis is not a random cut through a cell, like breaking up terrazzo flooring. As a form of asexual reproduction, mitosis qualifies as more-making because it duplicates an already coherent set of interdependencies.

Daughter cells of mitosis are more of an already organized coherent whole, the parent cell. Mitosis, that is, replicates the constraint regime of a particular type of entity, an organized cell. In transmitting bound type-level information from parent to an increasing number of daughters in subsequent generations, mitosis extends the reach of a distinct coherent constraint regime, a type. By duplicating an already coherent dynamic, mitosis prolongs an interactional type of entity into future generations.[2]

Cloning also makes another full organism like the original. Like mitosis, cloning replicates bound information by making more, more, more of the long-range correlations and constraint regimes that give coherence to a type. However, cloned animals like Dolly, the sheep, are born with the biological age of the stem cell from which they were cloned. Clones, in other words, replicate not only the parent's type identity; they also replicate at least some of the cloned organism's individuated characteristics. They carry the experiential record of the cloned organism's context-dependent constraints into a new embodiment. They transmit not only the inherited constraint regime but its epigenetic one. Clones are nevertheless replicates because, in addition to inherited constraints acting as constants, they embody a memory or record of the earlier ontogenetic path in their cloned structure and dynamics. Their very being embodies and extends a coherent trajectory. Replicating a differentiated cell type as well as an individuated token are characteristic of path-dependent processes operating simultaneously on different time scales.[3]

As instances of context-independent constraints, repetition, regularity, redundancy, and replication produce more tokens of the same type. None, however, directly increases type diversity or variety.[4] As noted in chapter 4, innumerable identical tokens of one type of entity are no variety at all. As obsessive-compulsive disorders demonstrate, diversity is not a cardinal measure. Cut off from contextual constraints that would ground them as meaningful, timely, and appropriate, repetitive habits and practices decoupled from context become compulsions.

This all changes with horizontal gene transfer (the lateral movement of genetic material between organisms, other than between an organism and its offspring) and the gene shuffling of sexual reproduction; these introduce novel context-dependent constraints that generate emergent interdependencies with an expanded possibility space. This new variety of context-dependent constraints does not replace earlier forms of more-making; it gets added on to them. Even *transposons*, sequences of DNA that jump from one site in the genome to another, jump to specific sites. They are context-dependent, not random.

Reproduction

Sexually reproducing organisms produce gametes (sperm and egg cells) through *meiosis*. Unlike daughter cells produced by mitosis, each of which carries the full information of the original, meiotic sequences yield four cells (sperms and eggs), each with half the chromosomes of one parent. These do not directly duplicate the full coherent organization of the original parent. Biologically, then, haploid germ cells are not replicates. Two haploid germ cells must combine to form an organism with the full complement of genes. Significantly, not only does the shuffling preserve the species' lineage, but individuals born of sexual reproduction also realize a novel combination of traits. New variants are produced, not by random mutations but directly by the multiply realizable constraints of gene shuffling; combined with the expanded possibility space afforded by horizontal gene transfer between bacteria, and between unicellular eukaryotes, a wider range of traits and behavior that nevertheless remains true to type opens up.

At the phenotype level, that is, sexual reproduction and horizontal gene transfer introduce new context-dependent constraints that can generate novel interactional types whose viability and life span are wider in scope and reach. These support the multigenerational constraint regimes we call lineages. They persist longer than their individual tokens. Constraints on both scales work in tandem. As a result, a population of sexually reproducing organisms has a more extended adaptive space in which to modify and survive (Kirschner and Gerhart 2005; Woods 2018). Subject to the limits of the possibility space set by the lineage's type, sexual reproduction increases flexibility and viability.

Phrased otherwise: gene shuffling, the preeminent context-dependent enabling constraint of sexual reproduction, does not make more of each parent, as mitosis does. By creating truly novel combinations (novel token interdependencies), gene shuffling breaks through the message variety trap of context-independent constraints. By endogenously producing novel variants, sexual reproduction creates a coordination pattern with a larger possibility space. Combined with horizontal gene transfer, these constraints enable the formation of multiply realizable diachronic interactional types we call biological lineages; these conserve unity of type while simultaneously allowing increasingly diverse tokens.

* * *

Epigenetic modifications and the gut microbiome's metagenome are further add-on enabling constraints that extend metastable coordination dynamics: symbiosis, mutualism, commensalism, and parasitism are variations of context dependence and coherence making. Epigenetics and the gut microbiome do not replace mitosis in individual organisms any more than sexual reproduction did; indeed, it can be difficult to determine which came first. The immunoprotection that the microbiome in the mother's birth canal confers on vaginally birthed infants is yet another example of interlocking context-dependent constraints enabled by feedback loops at more than one spatiotemporal scale (mother's and child's lifespans). Spatial contextual constraints change the state space's topology; they expand the coherent dynamics of interlocking constraints, extend constraint space, and allow its future exploration by actual organisms.

By increasing type diversity, reconfigured constraint regimes enhance their flexibility and, in consequence, their evolvability. Contextually constrained types, it will be recalled, are multiply realizable. Even today, some sexually reproducing species retain the potential to replicate asexually under certain conditions such as drought. Ordinary, temporally constrained transformations such as those from tadpole to frog, and from larvae to pupae to butterfly, are considered a single constraint regime that activates or inhibits processes in response to contextual cues. But these abrupt phase changes can also be considered distinct constraint regimes that alternate depending on cued contexts as much as different tokens of a persisting constraint regime.

The organism's genome recodes bound information as instructions for processes at multiple scales. As described earlier, instructions are sets of spatiotemporally coded constraints that realize and maintain tokens of a type. No phenotype displays the full range of its lineage's bound information; actual living things realize only a context-dependent subset of the lineage's more encompassing type-defining coherence (what Brooks and Wiley 1988 call H_{max}). Governing constraints modularized in an organism's genome preserve the species' type by controlling gene expression such that phenotypes remain within the boundaries of that constraint framework. Regulatory genes governing developmental processes are specialized to that function and cued to actual contextual features to which they were evolutionarily entrained and to which they differentially respond. By simultaneously satisfying a number of constraints, however, regulatory genes can transmit and conserve bound information that describes a circumscribed possibility space.

By increasingly relying on the enabling powers of context-dependent constraints such as gene shuffling, horizontal gene transfer, epistasis, and epigenetics (rather than on faithful more-making alone), the dynamic of coherence-making itself evolves. *Epistasis* makes the effect of a given mutation depend on the presence or absence of other genes and earlier mutations; epigenetics integrates environmental and behavioral influences without genetic mutations. Over evolutionary time, coherence making appears to increasingly incorporate context-dependent constraints like epistasis and epigenetics. As a result of these context-dependent constraints acting together, evolvability and natural selection are facilitated. It seems evident that by facilitating novel forms of interaction, this process of complexification accelerates and is likely responsible for the emergence of consciousness, symbolic thought and language, culture, industrialization, globalization, and other traits of the Anthropocene.

Unity of Type in Species, Demes, and Memes

"Species are strictly genealogical, evolutionary units." By way of contrast to "economic units," Eldredge's term for processes that transfer energy, species can be considered more-makers only if their historical constraints are effective, if they reproduce "heritable information ensconced in the germ line genome" (Eldredge 2015, 304–305). In the terminology used here, species realize a lineage's constraint regime by embodying and enacting heritable context-independent information modularized both in the germ line and in epigenetically and epistatically sedimented constraints regulating gene expression. Once again, the information that persists in each case is enacted as unity of type (Brooks 2010), a distinct second-order constraint regime that becomes increasingly generic and flexible over evolutionary time.

Such constrained interdependencies can also be transmitted and preserved across generations of social organization. *Memes* are constraints that establish boundaries, coordinates, and attractors in mental and social space. We can understand their dynamics better by thinking of them as socially transmitted constraints that frame and bias (weight) cognitive possibility spaces and their constraint architecture. They enable the reproduction and enactment of easily transmissible information about mental and social properties and parameters. The placement and traits of the mental and social attractors, including their ruggedness, change in response to the spread of memes.

Constraints enable memes and *demes*, subdivisions of populations that typically breed among themselves, to form and persist. No doubt geography served as the main enabling and restrictive constraint in the generation and conservation of demes, be they human, plant, or animal demes. Each set of coherent interdependencies that identifies a species, deme, or meme reflects a constraint regime turned context-independent module of conserved interdependencies. Such interlocking interdependencies are not monolithic and blocklike. By quickly adapting to each local culture, memes can shape-shift to individuate lineages, cultures, and mindsets. But by highlighting just one or two features, one cartoon or pithy comment can capture and easily convey the complexity of those interdependencies. "That's a Spartan lifestyle!"

Gatlin (1972) identified the key to vertebrate evolution as keeping basic constraints fixed while simultaneously allowing the influence of context-dependent constraints to expand dramatically. This theme and variation pattern seems to represent a general motif: enabling constraints precipitate a phase transition to a new possibility space, a constraint regime which itself becomes a new inhomogeneity that taps new gradients. Sexual reproduction, lineages, demes, and even memes are evolutionarily recent coherent dynamics that increasingly rely on context dependence to precipitate ever more extended and long-lasting interdependencies, with ever more complex tokens, in every more complex constraint spaces. Because all complex systems are constrained in path-dependent trajectories at multiple temporal scales, organisms become increasingly specified and individuated during their individual lifetimes. Epigenetics demonstrates that these modified and individuated traits can feed back into the dynamic and continue to influence it for some time.

Genes, species, and demes therefore qualify as more-makers. Each reproduces a distinctive generic governing constraint regime that persists as an identifiable and coherent type of entity, with characteristic parameters, and embodied and enacted in a wide range of path-dependent tokens. The constitutive and governing constraint regime of genes is embodied in nucleic acid; it differs from that of social demes, where it is embodied in culture and traditions, which in turn differs from those for memes whose central enabling constraint is the web. Different material substrates provide different constraints; the enabling constraints that generate each emergent trait are different; they may operate on a variety of spatiotemporal scales; their properties and powers can be qualitatively different. But in each case, context-dependent constraint regimes can evolve to become increasingly far-reaching and/or long-lasting; in turn, their realized tokens and phenotypes become uniquely individuated in response to the particular path they traverse, the local context that affects them, and even the timing of each step of the process.

Some governing constraints (like those of demes) may be more far-reaching than those of genes; their attractors are wider. By being more flexible and ambiguous, they span more tokens, at more levels of organization. Others, like memes, may be more contagious, and the slope of their attractors is shallower, making dispersion and transmission easier. The walls of the attractors of yet others are more porous and permeable to surrounding attractors. They resonate more with others. This is particularly true if, as suggested earlier, the internet serves as context-independent vasculature of memetic culture: its global reach and speed are ideal channels that facilitate easy adoption, adaptation, and better fit in different contexts. Its network architecture is a powerful enabler; as a result, memes can quickly sculpt dramatically shape-shifting attractors worldwide. Contagion models aim to capture the dynamics of that architecture, given a set of specified constraints.

In sum, by making more tokens of a coherent genetic lineage, sexual reproduction carries an existing coherent dynamic into the future. By making more tokens of a coherent deme or meme, social and cognitive constraints can prolong its constraint regime even longer. In each case, the interdependencies persist because increasingly flexible governing constraints exercise top-down control over actual cases while allowing local context-dependent constraints to individuate each token.

The rest of this chapter describes a number of other constraints.

Density, Isolation, and Buffers

In addition to gradients, catalysts, feedback, replication, redundancy, and reproduction, density, isolation, and buffers are constraints that also increase the stability of a system's existing default governing constraints and thereby prolong its existence.

Density alone promotes clumping. Like gravity, it does not directly produce irreversible correlations, interactions, and interdependencies the way catalysts, feedback, and recursion do. But proximity and packing due to density might have made it more likely that context-dependent constraints would appear. Earlier I speculated that, cosmically, a local gravitational well's density might have altered the likelihood that a new kind of constraint would arise. The fact that at the subatomic level forces and fields are interactions supports this conjecture. Constructive interference of nearby atomic orbitals (Nordholm and Bacskay 2020) constitutes a new set of interdependencies, a new possibility space: molecules with the novel powers and properties of chemistry might have been the chance consequence of dense gravitational conditions at the atomic level. Alone they would have imploded into a black hole, but when combined with the

Pauli exclusion principle, a constructive interference pattern that could act as a novel second-order top-down constraint might have emerged.

With fifteen billion neurons crammed into the rigid skull and 2.5 meters' worth of chromosomes packed into each cell, there are analogous reasons to hypothesize that crowded conditions in biological systems might have similarly favored the appearance of context-dependent constraints and generated new interdependencies among brain cells and with the environment.[5] Feedback from the environment through reentrant signaling in densely packed conditions, for example, might have correlated and intertwined neural processes. The resulting streamlining and pruning during early brain development in response to the infant's actions, body weight, and other contextual constraints offers evidence of such dynamics. Some of these constrained correlations might have then gone on to become modularized as distinct types of structural neurons, of which there are at least one thousand.

Isolation

Darwin's round-the-world trip on the *HMS Beagle* prompted him to focus, however, not on density and novel correlations due to close packing, but on the role of isolation in the formation of new species. Although it cannot retard genetic drift within a population, *isolation* stabilizes the status quo by preventing the introduction of new and possibly destabilizing information. As a context-independent constraint, isolation delays the homogenizing pressure of the second law by keeping conditions away from equilibrium longer than otherwise.

Darwin's genius was noticing that, long term, isolation can also occasion divergent clusters of interdependencies and new constraint regimes.

Isolation comes in diverse forms. It can be geographical. Populations in geographically isolated locations, like the Galapagos Islands, or in remote continents, such as Australia, diverged from other populations of (what might have previously constituted) the same species. Fish in different lakes can evolve into distinct species. Darwin's beloved finches are the textbook example of this process. In isolated conditions where the homogenizing role of genetic drift and gene shuffling is limited, selection of reproductively favored phenotypes (beak shape, size, and strength) will evolve into a distinct species whose interdependencies diverge from those of earlier populations. Geographic isolation can enable the formation of even more complex types.

Isolation and the accompanying species divergence keep the second law at bay longer than expected. In contrast, interbreeding brought about by isolation paves the way for the reemergence of *latent constraints* in recessive genes. When contextual changes weaken existing governing constraints, incestuous breeding that would not occur under normal conditions might hasten the appearance of recessive traits. Possible realizations that were dormant resurface. The process, however, depends on the fact that the species, as a coherent type, is multiply realizable and variously embodied in actual tokens depending on contextual constraints.

Both species divergence and the resurfacing of latent traits appear to be evidence of an increasing role for context dependence over evolutionary time. The distinction between dominance and latency is context dependent; it is conditioned on an embedding context with respect to which the genes in question are dominant or recessive, manifest or latent. Latency, that is, is defined with respect to an existing contextually constrained type induced not only by the enabling constraints that generate it and the governing constraints that preserve its coherence, but also by the current context within which those constraints effectively tailor and modify outcomes. Dominance and latency in biology therefore enact *hybrid constraints*, which combine context-independent and context-dependent characteristics. Imposing or relaxing such constraints changes the landscape of conditional probabilities and therefore the events within it.[6] Because recessive is type-defined, and types, in turn, are contextually generated, recessive phenotypes surface or not depending on circumstances. Under conditions of reproductive isolation, the reappearance of latent constraints reflects gene shuffling statistics that, in a less diverse possibility space, make it more likely that recessive phenotypes will reappear. Of those that do, fitting in with the new conditions will be adaptive; the stressed type, too, will correspondingly evolve.

In isolated populations, resurfaced latent constraints and horizontal gene transfer can produce mixed outcomes. Some of these recessive phenotypes might have survival advantage and will be selected. But a higher prevalence of autosomal recessive disorders directly attributable to isolation and interbreeding is also more likely. Communities such as the culturally and geographically isolated Old Order Amish exemplify this dynamic. Changing the ratio of dominant to recessive genes by altered constraints can place unity of type at risk.

The central point here is that conditions in which latent constraints reappear and isolated species diverge need not be the direct consequence of random mutations alone. Genetic as well as epigenetic and epistatic information is encoded in multiply realizable and constrained interdependencies at numerous spatiotemporal scales; these establish conditional probabilities for gene activation, inhibition, expression, and so on—always conditional on context. Under certain conditions, those with a certain likelihood are dominant; under those same conditions, those marked by a different probability are recessive. Under entirely different conditions completely new phenotypes might appear without genetic modification. Actual phenotypes that are not metastable and fail to fit the context, of course, will not reinforce, repeat, or reproduce. The niche's overarching governing constraints will not embrace them. But the capacity to span, adapt, and surface a range of variants allows constraint regimes themselves to coevolve in tandem with the extant realizations. In a multidimensional state space, the combinatorics of multiple constraint satisfaction can be mindboggling.

As the human genome project confirmed, these ideas imply that type–token relationships are not one to one. Multiple realizability of types, abiotic and biotic alike, implies a range of variant tokens that are actualized or not conditional on context; it implies that interactional types span multiple paths to realizing the same emergent and collective properties.

Analogous dynamics also accelerate divergence in populations that are not geographically isolated. Differentiation keeps state space inhomogeneous, but differentiation due to isolation can come about due to ecological reasons. Food preferences can cause insects to feed in distinct types of trees. Over time, those insect populations become distinct. Similarly, differences in mating preferences even in mixed populations drive changes in the probability distribution of the various phenotypes. Distinct habitats form. Over time, gene pools diverge and produce "genomic islands of divergence" (Turner et al. 2005).

Recent research on two species of birds that share the same nature preserve in Argentina confirmed sexual selection (Campagna and Turbek 2021; Turbek et al. 2021). These two species of seedeater differ in plumage color and the details of their songs. Sharing 99.9 percent of their genes, brightly colored and tawny seedeaters interbreed freely in captivity. But they never interbreed in the wild. If they did, gene shuffling would homogenize their traits and prevent divergence. Whereas unconstrained sexual reproduction homogenizes a population, mating preferences and sexual selection are context-dependent constraints that change the proportion of latency to dominance. The evolution of mating preferences and sexual selection, in other words, indicates the emergence of a new context-dependent constraint. Once established, that new constraint regime is made manifest as new local troughs and hillocks in possibility space.

One sort of isolation can contribute to another: *geographic*, *cultural*, *religious*, or *informational isolation* can significantly restrict the reproductive pool from which mating preference selects. In the case of the Old Order Amish, the effective upshot of combining these various forms of isolation is *reproductive isolation*. Reproductive isolation brought about by mating preferences and other forms of sexual selection can also sediment and entrench differences, as will be discussed below.

As a reminder: these remarks are pertinent to our central concern, coherence-making, for several reasons: (1) neither mating preferences, isolation, density, nor sedimentation and entrenchment, as will be described below, are efficient causes, and yet they have consequences. They are constraints that generate, preserve, modify, and realize interlocking interdependencies. (2) They have consequences as constraints depending on the spatiotemporal context in which the entity in question is enmeshed and embedded. Even more to the point: (3) the above dynamics are possible only because the governing constraints of coherent interdependencies are multiply realizable and were generated by enabling context-dependent constraints.

Shielding organisms from gene shuffling and genetic drift generally results in species divergence. In addition to isolation, other constraints acting as shields include buffers and boundaries.

Buffers and Other Shields

For open systems that exchange matter, energy, and/or information with their environment, preserving coherence over time—prolonging a complex system's existence—is a question of sustaining their constitutive and governing constraint regimes despite and throughout those exchanges. As has been repeatedly noted, constitutive constraints stabilize interdependencies in the face of fluctuations and perturbations; they carry out that stabilization as governing constraints.

By adding extra context-independent constraints, buffers further stabilize existing constraints and delay thermalization.

Buffers are conditions that alter possibility space such that energy flows more readily in one direction than another—or does not flow at all. Buffers are not forces; they are add-on constraints that reinforce existing governing constraints by shielding complex systems from intrusion by extraneous forces. Buffers prevent potentially destabilizing elements, events, or information from gaining access. More precisely, buffers are impediments to energy flow. Indirectly, they strengthen boundaries and interfaces and thereby improve the likelihood of a coherent system's continued existence. Chemical buffers, for example, dilute the effects of threats to the system's coherence. Buffers are thus extra constraints that, under precarious conditions, help preserve the status quo.

Buffers need not be physical. Rate differences between a system and the wider context in which it is embedded can also dynamically buffer systems from their environment. In ecosystems, buffer species are distinguished by different cycling rates than other species—and the environment in which they exist. If predators can feed more readily on buffer species, either because there are more of them than of the usual prey, because they reproduce faster than the usual prey, or if there is a scarcity of the usual prey, rate differences protect the ecosystem's overall balance. High reproduction rates also buffer against population crashes by adding a context-independent constraint that allows lineages to temporarily outrun entropy. Again, rate differences are not efficient causes; they are constraints that generate and constitute distinct coherent dynamics.

Buffers can also be temporal. Audio lag time during live radio, television, and even Zoom calls prevents inappropriate or potentially illegal comments from being broadcast. Likewise, buffers need not be exclusively spatial. As we saw in the section on isolation, geographic isolation initially acts as a spatial buffer and brings with it its own constrained interactions; it can then do double duty as an informational buffer. But direct informational constraints such as censorship and limited access to sources of financial support or information also buffer the status quo from novelty that might destabilize the system's informational closure.[7]

Buffers can emerge spontaneously in response to context-dependent constraints. Early agent-based models of urban dynamics showed how neighborhood segregation naturally arises from interactions constrained in particular ways (Schelling 1971; Hatna and Benenson 2012). Buffers are also common in computer coding, where data are buffered in a separate storage area for future editing. In chemistry, liquid–liquid phase separations demonstrate the self-organization of buffers. Some chemical solutions also maintain their pH relatively stable by buffering against pH change. Stable pH conditions serve as a background constraint that allows enzymes that require precise pH levels to function.

The distinction between buffers and boundaries is hard to draw for open systems. The term buffer is traditionally used to refer to separations caused by material substrates. Expanses of ocean buffer the Galapagos from invasive species. Mountain ranges and water obstacles (including artificial moats) are called buffers, as is the demilitarized zone on the Korean Peninsula. But buffers need not be materially based. Zoning regulations and financial redlining practices can deliberately include construction and financing rules that keep races separate and thereby consolidate segregation.

To summarize: isolation and buffers are constraints that support persistence by protecting a system's zone of metastability and continued existence, which they do by acting as shields against threatening input. Isolation and buffers bring about this added protection through context-independent constraints that guard existing coordination dynamics from environmental threats that might precipitate their dissipation. Feudal lords did the same with castle keeps and moats.

Once again, it is tempting to conjecture that the universe is characterized by a general trend, from rigid boundaries and physical gaps to interactive interfaces operating as *gating constraints*. Just as rigid walls of prokaryotes gave way to more permeable membranes of eukaryotes, so too moats and Great Walls of China gave way to . . . safe conduct letters of transit and visas? In each case there emerged dynamic interfaces that actively negotiate relations between the inside and the outside and thereby satisfy multiple constraints. If this conjecture were confirmed, it would be worthwhile to study the role context-dependent constraints played in that trend. Increasing complexity suggests increasing reliance on context dependence.

Motility and Migration—The Relaxation of Constraints

Human and animal populations as well as convection cells can exhaust or deplete existing energy gradients that nourish them and which they tap to survive or persist. Such risks to enabling and governing constraints of existing coherent dynamics can destabilize their effectiveness; interdependencies begin to decohere and disintegrate. Entire metastable dynamics must adapt and evolve in the face of such threats; otherwise, extant governing constraints will inevitably weaken. Invasive species whose constraint regime offers a significantly better fit with current conditions can outcompete autochthonous but threatened ecosystems. They do not become buffer species; they wipe out native ones.

In contradistinction to reinforcing existing constraints, relaxing them might open the door to new enabling constraints. Motility, for example, opened[8] new degrees of freedom when existing and restrictive constraints that kept plants in a fixed location were relaxed. Relaxing those constraints was tantamount to making room for the appearance of new enabling constraints. Affixed to a location, plants change their shape throughout their existence. Animals, in contrast, trade shape shifting for the ability to change location; as adults, their shape changes little. Different enabling constraints induce different interlocking constraints, which realize different forms of life. Beginning with chemotaxis, cellular locomotion is a novel form of life induced by the resulting phase transition enabled when barriers to energy flow were relaxed. New options emerged: moving toward a higher concentration of a diffusible substance or away from it. The appearance of motility in living things became a new enabling constraint that opened possibility spaces with previously untappable free energy. It changed things completely.

In response to other context-dependent constraints, some animals take motility one step further; they become migratory species, thereby expanding the spatiotemporal reach of their constraint regime. Migration can be either a one-off phenomenon or seasonal. The latter is indicative of expanded governing constraints cued to context-dependent signals such as ambient temperature or light, or the availability of resources. In response to gradients, migrating populations gravitate toward more favorable environments when their existing niche is threatened (Agosta and Brooks 2020).

From hominins over two million years ago to human migrations today, mass movement patterns are evidence of resource gradients operating as context-independent constraints. Because they are embedded in an extended ecosystem dynamic, populations respond to contextual cues and, as a result, can precipitate mass migrations. They move toward improved resources and away from threats. Events in the U.S. southern border illustrate how stressful conditions at home, combined with the perception of accessible opportunity in the United States, provide multiple constraints that take a system far from equilibrium to a threshold of instability and then over it to new forms of interdependence.

Migration is directed toward adjacent possible areas where there are (or are perceived to be) available resources or fewer threats. All these terms must be understood as referring to spatiotemporal constraints and ultimately cashed out by thermodynamic considerations. In an era of airplanes and now space tourism, even the term *adjacent* must be redefined. People will migrate to areas where they know others, often from the same village. Here, adjacent translates into known and culturally compatible overlap, not physical distance. The general point here is that contextual considerations contribute to outward as well as internal migration by relaxing and reconfiguring existing enabling constraints, including overall constraint regimes and the topology of possibility space. Top down, this reconfiguration changes the beliefs, dispositions, and behavior of individuals caught up in that dynamic.

Outmigration renews a population's possibility space. Once a population migrates, its constraint regime imports close-enough features of the new and previously unfamiliar habitat, whose own constraint regime will simultaneously intrude into the population's earlier constraint regime and thereby co-create a wider niche or adaptive space (Woods 2019). Both coevolve. The transformation of the sociocultural landscape of the southwestern U.S. border, or of Miami-Dade County (Florida), over the past sixty years in response to migration flows provides a dramatic case study of this process. The oft-heard phrase, "Miami is the capital of Latin America," implies a wider set of financial and cultural interdependencies whose governing, top-down constraints (many of which are tacit) are affordances of the U.S. legal and political system. Its fiscal and jurisprudential framework affords the metastability and overlapping habitat necessary for hemispheric trade relations. But both habitat and residents coadapt as a result.

It is not too far-fetched to propose that the translational motion of hurricanes and the oscillating waves of BZ reactions are physical and chemical precursors to animal motility in general and human and animal migration in particular. In contrast to their rotational motion, the forward, translational motion of hurricanes is sometimes called propagation, making the analogy with disease spread explicit. This is possible because the paths of migration, epidemics, and hurricanes are not random walks. They are enabled and governed by multiple constraints. The direction and speed of cyclones and hurricanes are vectorial coordination dynamics dependent on constraints; these generate a tendency to move toward warm areas and away from frigid air. Likewise, epidemics of infectious diseases will spread to areas with increased population density and low rates of inoculation or immunity, for example. People move to areas where they know others, or of safer conditions. The upshot, once again, is that enabling constraints generate patterns of matter, energy, and information flow; governing constraints then preserve the coherence and persistence of those patterns by exploring/exploiting previously inaccessible regions of adaptive space. In turn, the dynamics expand spatiotemporally and last longer than they would in the absence of new coherent interdependencies or tacked-on constraints such as buffers, isolation, and the capacity to move.

This reinforces the point that niches and habitats are not existing sites, prebuilt recesses in a wall waiting for fit organisms to show up and occupy them. Niches and habitats—abiotic, natural, and social—are co-constructed contexts; they are coordination dynamics interwoven by constraints. Those interdependencies manifest as a contextually coherent type that underpins novel affordances for multiply realized tokens (Gibson 1975; Carello and Turvey 2002; Chemero 2003, 2009). Those interdependencies then select for actualization from among the range of options afforded by the new interactional type.

Failure to understand the dynamics of contextually constrained niche formation and modification will get a lot of the actual interdependencies wrong. Failure to appreciate the background constraints of entrenched economic, cultural, and political conditions against which context-dependent constraints operate will get yet other interdependencies wrong. Failure to recognize tacit constraints risks bottom-up upheavals when governing constraints are unresponsive to context. As always, enabling and governing constraints intertwine simultaneously and over different scales, requiring multiple constraint satisfaction. Failure to track underlying constraints responsible for the invariances gets the ontology wrong. If it is to have any hope of being effective, domestic and foreign policy must get these social constraint dynamics right.

Templates, Frameworks, Scaffolds, and Affordances

The concepts of scaffolds, frameworks, templates, and affordances highlight important forms of constraint (Wimsatt 2007; Caporael, Griesemer, and Wimsatt 2014; Bickhard 1992; Carello and Turvey 2002). They describe specific varieties of constraints in social and biological settings. Some are context dependent; some are not. Some are enabling constraints; some are governing constraint regimes. All describe varieties of influence other than mechanical or forceful causes. They are constraints that inform cognitive and behavioral possibility spaces and guide realizations to completion. They do so by presetting and then continuously adjusting the likelihood of what is possible and what is likely.

Affordance theory (Gibson 1975; Turvey, Shockley, and Carello 1999; Chemero 2003) was among the first to view psychological capabilities in ecological terms. *Affordances* are relational properties that function as constraints by weaving together an organism's capabilities and propensities and its world. Affordances are the products of enabling constraints that create novel interdependencies and opportunities. Like biological niches, they structure an organism's psychological ecology, dispositions, and behavior patterns such that it can act more effectively in that world. To use a common example, for Westerners, chairs afford opportunities for sitting. For the Japanese, tatami mats afford opportunities for sleeping. Individuals embedded in those worlds recognize those items as affordances and act appropriately (Chemero 2003. Also see chapter 13 of this volume, the 4E approach).

Because they are formed by context dependent and ecological interdependencies, affordances can activate in sequence in response to other temporal and contextual constraints. Walking is not afforded to newborns, but it is to toddlers (Thelen and Smith 1994). Differentially activated constraints make possibility spaces even more complex and effective. Superior design is all about incorporating transparent affordances and enabling constraints into products and services, from software and business development to commercial and industrial artifacts and communities of practice generally (Bloom 2015; Burgauer 2022b). Under the guidance of English American experimental psychologist Michael Turvey, UConn's CESPA lab used the concept of affordances to design prosthetic devices that build in enabling constraints to improve perception and action.

* * *

The *Oxford English Dictionary* (*OED*) (June 2022) states that, used metaphorically, the concept of *framework* suggests the idea of "fitted arrangements." In this sense, frameworks are like affordances and principles of conceptual organization; they format behavior by presetting definitions, parameters, and category boundaries—in short, by initializing cognitive and behavioral constraint spaces within which certain ideas and actions will be possible or viable, and others not. Conceptual frameworks allow room for adjustments and arrangements before the fact; not so much afterward. Like IT platforms, conceptual frameworks initialize rules and parameters—the background constants—on which instructions and protocols rest. In short, frameworks are constitutive and governing constraints that structure, coordinate, standardize, and otherwise adjust relations among the concepts and assumptions they organize. As a result, these organize coherent conceptual landscapes.

Metaphorically speaking, frameworks are looser than molds and templates. Because molds and templates guide behavior to ensure a certain outcome, they incorporate motivational constraints, which frameworks do not, at least directly. Templates actively direct behavior while it occurs; they also limit the possibility of changing the template itself once the process has begun. Chapter 7 described how folding back onto itself creates templates for protein folding. Templates operate throughout the biotic world. Prions (proteins, initially thought to cause only harmful effects such as the neurological disorders Creutzfeldt-Jakob disease and mad cow disease) have recently been found to empirically implement template formation. Under certain conditions, a protein called CPEB folds back into a prion, which then grows a fibril that permanently alters synaptic structure. Described as *self-templating*, this constraint-directed process is hypothesized to be responsible for a protein-based inheritance mechanism—a form of long-term memory that is independent of DNA (Brahic 2021).

Scaffolding

Scaffolding has recently become the object of attention of practitioners in fields as unrelated as theoretical biology and business management. American architect Ann Pendleton-Jullian (2020, 280) maintains that scaffolds are a more generalizable concept than frameworks. Both, however, inform. The following section draws heavily from foundational work on scaffolding by Wimsatt (2007); Wimsatt (2014) in Caporeal, Griesemer, and Wimsatt (2014); and Bickhard (1992).

Pioneered by Soviet social constructivist Lev Vygotsky and developed by American cognitive psychologist Jerome Bruner, the metaphor of instructional scaffolding was first applied to structured interactions between teacher and pupil to guide learning. The activity of scaffolding is a context-dependent constraint that, catalyst-like, facilitates information, matter, and energy flow—in a certain location, in a certain direction. Teacher-provided scaffolds offer points of equilibria and pathways that make the next step easier, smoother, or quicker for the pupil. In the formation phase, then, scaffolds operate as enabling constraints that assist in taking projects to completion. Scaffolds set up affordances and offer nearby points of equilibrium (Bickhard 1992) or "zones of proximal development" (Vygotsky 1978) that stabilize progress and make next steps more accessible.

The stabilizing power of scaffolds is key. Scaffolding's simultaneously enabling and stabilizing constraints make it more likely that goal-directed activity will reach conclusion. Scaffolding is therefore critical to the successful completion of many complex projects. Without them, the project might not be finished; it might not even get off the ground.

Whether taken as metaphorical or literally, *ratchets* are also scaffolding mechanisms. The asymmetry of a ratchet's gear teeth preserves and stabilizes each achieved step. By preventing backsliding and allowing movement only in the forward direction, ratchets make it easier to reach the next step. Ratchets and scaffolds thus reinforce sequential enabling constraints that make advancing easier by providing a steadily growing platform (Bickhard 1992) from which to launch the next step.

By affording nearby "points of stability" from which to continue building or growing in the right direction and at the right pace (Bickhard 1992), scaffolds and ratchets are also like affordances; they embody context-dependent constraints that bias the likelihood of which future events become possible, can occur more readily, or be performed or structured in a particular way. They set up interdependencies whose joint probabilities prevent disruptive perturbations from waylaying long-term projects. In doing so, they contribute to the formation and persistence of new interdependencies. They contribute to coherence-making.

By characterizing scaffolds as "muting, blocking, or suspending selection pressures," Bickhard 1992 implies an ecological perspective. Scaffolds afford guidance and progress only for those who are cued to the opportunities of a particular context and artifact. As mutually constraining and constrained processes, scaffolds are therefore afforded by the shared interdependent constraint regime in which both agent and scaffold operate. The adaptive space of teacher and student, for example, is already socially structured even before the two walk into the classroom. Similarly, Asia's bamboo scaffolds in skyscraper construction take advantage of a plentiful and renewal resource, but only if builders are attuned to the requirement that bamboo must be pretreated against fungi and insects to function as scaffolds at all.

Scaffolds and affordances therefore facilitate natural and cultural evolution; they make stepwise progress easier by preselecting and biasing the user's next steps. Cultural rituals and taboos do the same for members of a community; mourning and grieving traditions such as sitting shiva and ceremonial wakes no doubt serve that function as well. By weighting the possibility landscape, preset constraints allow members of a community not to rethink each step on their own. Coherence-preservation is facilitated as a result.

When a coherent dynamic's governing constraints weaken, its gear teeth destabilize and become unable to prevent it from degrading into thermal equilibrium. This understanding of context-dependent constraints is congruent with Bickhard's statement that artificial tools and even entire cultural environments can be considered scaffolds that facilitate the construction of "a structure distributed in space" (Bickhard 1992); such constraints are affordances, "adjacents possible" that lower barriers to energy flow and thereby make next steps more accessible. Scaffolds and scaffolding, molds and templates are context-dependent constraints with ratchet-like properties.

By including temporal constraints, we can extend Bickhard's insights to include scaffolds that aid completion of "structures distributed in time," such as the constraints of ontogenetic development and lineages. (For differentially activated constraints, see generative entrenchment in chapter 10.) From regulatory genes to yearly family rituals or once-in-a-lifetime ones like baptism, temporal constraints generally provide time-based scaffolding and affordances that help complete processes and keep social dynamics coherent. Recall the precise sequence of steps required to prepare cassava root and nardoo spores. Such metastable sequences are temporally coded sequences that, as scaffolds, lessen the effects of external insults; they also minimize having to figure out everything from scratch each time. In consequence, they mute selection pressures. Entrenched constraints stabilize those existing metastable sequences and dynamics. They also facilitate the appearance of other enabling constraints that generate major transformations. Entrenchment and modularity will be explored in more detail in chapters 10 and 12, respectively.

Types of Scaffolds

Like the term constraint, of which scaffolds are one realization, *scaffold* itself is ambiguous. It can refer to either a process or a result. The classical notion of scaffolds, of course, is *artifact scaffolding*, such as traditional architectural scaffolds, those externally manufactured and assembled objects that facilitate the construction of tall buildings. They bias the flow of energy and matter in a particular direction, in a particular sequence. *Architectural scaffolds* are thus prebuilt context-independent constraints that take the possibility space away from equiprobability. Following Bickhard, national monuments and museum collections of material culture can also be considered scaffolds that promote cultural transmission. They encourage us to see history in a particular way.

Asking the right questions in the right order at the right time, on the other hand, is a *pedagogical scaffold* that does not produce an artifact. Other varieties include *infrastructure scaffolding*, such as transportation and utility networks, and *developmental scaffolding*, such as the Peace Corps and Teach for America organizations. Some of these produce an artifact; others do not. Capacity building is a combination of infrastructure and developmental scaffolding.

Cellular cytoskeletons and vertebrate backbones, too, are context-dependent organic scaffolds. They provide stable, and therefore more durable, spatial configurations that serve as frames for the construction of organic structures. Their structural soundness and configurational stability generate novel (vectorial?) enabling constraints that guide processes along the correct path on which to build or repair further stable structures. In the abiotic domain, metastability likewise prolongs constraint regimes and thereby allows future increases in complexity to occur.

For purposes of this book, then, scaffolding, scaffolds, affordances, frameworks, templates, and molds are significant for their role as constraints. They facilitate, bias, and guide behavior—toward directed goals or end points, along precise sequences of stable steps. They embody implicit instructions, sequences of stepwise spatiotemporal constraints. Under precarious conditions, scaffolding also helps ongoing developmental processes avoid dissolution and successfully reach finality or even get started. Once set up, scaffolds serve as background constants of possibility space. Scaffolds and scaffolding are thus hybrids of context-independent and context-dependent constraints.

The metaphorical use of the term scaffolding is commonly organized around the workings of architectural scaffolding. In consequence, it is often characterized as temporal and temporary resources that provide a platform from which to ratchet development generally, not only learning (Caporael, Griesemer, and Wimsatt 2014, 2). However, thinking of scaffolds generally as temporal and temporary (as is the case in architectural and learning scaffolding) is limiting. Not all scaffolds (as enabling constraints) are temporal or temporary; even external architectural scaffolds can become integrated into the structure they helped construct. Flying buttresses of Gothic cathedrals are one such example.

Cynefin founder David Snowden and architect Pendleton-Jullian (2020) identify three additional types of scaffolding: keystone scaffolds, neural scaffolds, and nutrient lattices.

Arch construction is enabled by a particular form of architectural scaffolding, the keystone. Once in place, *keystones* become a permanent point of stability from which further construction can proceed. They become, in other words, permanent scaffolds for future growth. *Nutrient lattices*, on the other hand, are temporary. Applied to burn wounds in skin grafts, nutrients embedded in the lattices promote collagen production. As the body produces more collagen, "the connective tissue works its way up the artificial scaffolding, slowly building up new dermis. Over time, the artificial scaffolding dissolves away leaving no trace of the lattice."[9] This is an example of efficient causes (nutrients and metabolism) guided by scaffolds or affordance-like constraints (lattices and collagen).

Scaffolding that starts out as an external artifact and boundary condition can also become incorporated into the system itself without dissolving. Sometimes it is even integrated into the structure itself and becomes an internal constraint that confers stabilizing oomph and thereby contributes to the structure's persistence. For example, artificial bioactive lattices implanted to encourage bone growth do so as porous templates. As new growth infiltrates the interstices, the artificial lattice becomes permanently incorporated into new bone (Pendleton-Jullian and Brown 2016). The location, orientation, and size of a lattice's pores—their arrangement—embody its template function; these are enabling constraints. Once woven into new bone, they become components of the constitutive/governing constraints of the skeletal system.

Boundaries between scaffold and system, between system and environs, and between constraints and overall niche reflect differences in constrained interdependencies. As the prairie grasslands example illustrated, when feedback loops extrude into the environment or, which amounts to the same thing, import the environment into the dynamics, the dichotomy of internal versus external becomes irrelevant to any definition of scaffolding. It is the informational import of scaffolds—the constitutive or governing constraints that emerge from interwoven enabling constraints—that constitutes the effective interface between scaffold and scaffolded. It initializes the gating criteria of that constraint space as well as the settings that control action.

Whereas scaffolds enable process completion, prosthetic devices such as crutches, hearing aids, and artificial limbs as well as prescription eyeglasses and implanted lenses contribute to partial or full restoration of function. They are *restorative enabling constraints*. Their design affords links whereby emergent properties such as locomotion, vision, and hearing can be reestablished. What begins as a context-independent constraint becomes a context-dependent constraint that integrates new pathways into the coherent dynamics and constraint regime that embody a function. Implanted, *neuromorphic prostheses* (a variety of *neural scaffold*) that incorporate machine learning can become, with training and use, second nature, their deployment as effortless as the organic limb they replace.[10] Nowadays, brain computer interfaces (sometimes referred to as "brain–machine interfaces," or BMIs, when the signals are recorded by implanted sensors) can control a robot arm in real time. *Stentrodes,* electrode arrays inserted into blood vessels in the brain, monitor neural activity in part of the cortex. In time, the information conveyed by those implants becomes seamlessly incorporated into the subject's second-order, top-down constraint regime of motor activation: by thinking words or sentences that are then transcribed to a computer, patients can email, use apps, do online banking, and even turn the lights on and off in their house.

Alternating between *external* and *internal scaffolding* is a common pattern in nature. The transition from external to internal can be a form of coopting, as when external constraints are brought onboard and their function is internalized. Moreno and Mossio (2016) characterize the process of internalizing or coopting previously external boundary conditions as shanghaiing them and incorporating them as internal ones. When mitochondria were engulfed into bacteria, the former's enabling constraints for energy production were effectively coopted. The resulting interdependencies and new governing constraints became a new type of organism, eukaryotes. In the process, the rigid wall of the earlier prokaryotes cells also transformed into the eukaryotes' flexible but structurally integral combination of permeable cell membrane plus endoplasmic reticulum and cytoskeleton—an internal scaffold. As an active interface, this new structure is a constraint structure consisting of multifunctional microtubules, intracellular filaments that shape the cell, assist in intracellular transport, and even guide mitosis to ensure the cell's coherent structure is correctly replicated. In effect, the microtubules have become internalized scaffold constraints that prevent the cell's internal but mutually dependent components from equilibrating. As a curious aside: images of cell cytoskeletons are remarkably similar to images of the webs of cosmic dark matter mentioned earlier that appear to provide an invisible backbone that determines the hidden structure of the cosmos (Kruesi 2013).

We can suppose that, in analogous fashion, the external scaffolding of invertebrate exoskeletons (think marine seashells and horseshoe crabs) was shanghaied and evolved into the internal scaffolding of vertebrates, organisms with a more complex constitutive constraint regime. Even regulatory genes today are hypothesized to be modular encapsulations of what might have earlier been external constraints.

Once again, it is not entirely unreasonable to conjecture that such cooptation and shanghaiing of external constraints might represent a general trend that goes from externally set and rigid and impermeable boundary conditions of Bénard cells to endogenously generated and more malleable, porous, and internalized constraints whose most recent manifestation is the emergence of self-constraint in living things.

If so, closure of constraints in living things, as described earlier, constitutes the emergence of a form of *self-scaffolding*.

More-making makes more tokens of a coherent type, but closure of constraints ratchets the process to one that *self-scaffolds as self-constraint*. Closure of constraints self-constrains, self-scaffolds, self-constructs, and self-maintains interdependent processes in dynamic but steady nonequilibrium. Moreno and Mossio (2016) note that the being of living things "is their doing." The interweaving of these various forms of constraint integrates and preserves the interdependencies that define living things as the self-determining and autonomous variety of self-organizing process.

9

Persistence—Delaying the Second Law

Persistence can be described as stability over time (Pascal and Pross 2015). As realized in biological lineages and cultures, character traits, family resemblances, and social practices, constraint regimes endure longer than their moment to moment realizations. What persists are interlocking and covarying mutual dependencies held together by constitutive and governing constraints across spatial and temporal scales. This persistence holds even if the possibility landscapes that contain those interdependencies turn increasingly rugged over the entity's existence or lifespan. This rumpling of possibility space occurs in response to other more local and timely constraints. Earlier in this work, coherent interdependencies were referred to as realizing unity of type. What persists in individual realizations is not the material substrate of concrete particulars but the stored information embodied in constraint regimes.

In the case of living things, that metastable information enacts inherited genomic, epistatic, and epigenomic constraints. It is carried by the enabling and constitutive constraints that hold together as a configuration in dynamic metastability. Earth's metastability was generated and is preserved by a constraint regime that persists despite significant changes in its constituent details, such as the introduction of photosynthesis. Dissipative structures are also metastable, and for the same reason: enabling constraints in open systems under conditions far from equilibrium first precipitate a phase transition; the interlocking constraints of the resulting constitutive regime then preserve its coherence despite perturbations and fluctuations. Previous chapters emphasized that such contextually constrained interactional types are not universal, eternal, and unchanging; they are local and temporary. But given the contextual constraints that generated them, and while their constitutive and governing constraints hold, they are real and effective. And they persist.

Which sorts of constraints generate persistence? Previous chapters described processes and constraints that, upon reaching closure, interweave mereological relations. We saw that the overarching hyperloop formed in constraint closure takes longer to complete than the individual reactions that constitute it. By folding reactions back on themselves, recursion, autocatalysis, and feedback loops reinforce, replicate, and reproduce those conditions, components, and reactions that preserve mereological relations and prevent decoherence. They also incorporate real features of their world into their very constraint regimes.

This chapter focuses on persistence as a product of temporal constraints. Chapter 10 surveys diverse forms of entrenchment, a temporal constraint that reinforces persistence.

Persistence and Thermodynamics

How do coherence and persistence fit in with thermodynamics?

As articulated by Clausius, the second law of thermodynamics states that it is impossible for a process only to transfer heat from a colder to a warmer body. The end point of any thermic process is a state of equilibrium with its environment, a permanent and unchanging condition.

The received interpretation of the second law as formulated by Austrian physicist and philosopher Ludwig Boltzmann rests on a distinction between macro- and microstates. *Macrostates* are descriptions of large numbers of identical and independent particles, like molecules of gas. The temperature and pressure of a thermodynamic system, for example, can be realized by more than one microstate, the position and velocity of each molecule of gas.

The second law's arrow of time describes probabilistic tendencies of macrostates. Consider gas molecules in an enclosed container divided into two reservoirs, one hot, the other cold. Such a highly improbable initial condition (most molecules confined to one end and few molecules in the rest) is the system's initial macrostate. It lasts only while the partition between the two sides remains secure, or energy continues to be pumped in to keep the two sections separate. Once the partition is removed, the existing gradient inexorably dissipates.

Macrostates like this one that are realized by fewer configurations (two reservoirs) are less likely to occur than those with more configurations. The textbook heuristic is the following: throwing a pair of six-sided dice will come up a total of 7 more often in the long run because more combinations realize macrostate 7 than other macrostates. Despite each throw of the dice being an independent event, long runs of double sixes or snake eyes are possible. Unlikely, but possible.

Macrostates with a more uniform distribution of microstates are more likely than inhomogeneous ones such as the initial condition where molecules are bunched over at one end of the container and a near vacuum at the other. Over time a tendency appears, from less likely (more inhomogeneous) macrostates to more likely (more uniform) ones.[1] To illustrate: as more throws of the dice occur, 7 appears more often than double 6s or double 1s. In thermodynamic processes the distribution of molecules becomes increasingly uniform until it reaches the inevitable (most likely because most uniform, and therefore most stable) state of thermal equilibrium. As required by the second law, the tendency's overall direction, from least likely to most likely, is irreversible until the process reaches thermal equilibrium, the equivalent of white noise.

A tendency is a probabilistic measure. The statistical interpretation of the second law therefore equates *more likely* (macrostate probability) with *more stable*. And most stable with unchanging and therefore most lasting. Decades before Richard Dawkins repeated the quip, Herbert Simon (1969) riffed that this interpretation of the second law amounts to "the survival of the stable."

What, exactly, does this lasting condition amount to? Once reached, thermal equilibrium lasts, but there is *no-thing* that persists at thermal equilibrium (Ladyman and Ross 2010). White noise is no-thing; equiprobability is no-macrostate. Macrostates that appear to pop up along the way to thermal equilibrium (like a long run of snake eyes) are statistical improbabilities. Since on this interpretation microstates are independent of each other, those macrostates do not represent covarying relations formed and preserved by interlocking context-dependent constraints. Thermal equilibrium at the end lasts because the total absence of constraints ensures permanent no-thingness. The path to thermal equilibrium is the process of progressive relaxation of arbitrarily and extraneously set context-independent constraints (Barbour 2020).

Riffing on Simon, the second law could therefore also be characterized as "the survival of no-thing" since at thermal equilibrium there are no constraints and consequently no stored information that is qualitatively other than the aggregate or sum of the equally likely *microstates*. There is no stored information because there are no covarying interdependencies. There are no covarying interdependencies because there are no constraints. Thermal equilibrium carries no information and can do no work. Unconstrained equilibrium (white noise) is a statistical description, but it is a macrostate in name only. There is no "it" that persists as itself over time (Ladyman and Ross 2010). Thermal equilibrium is not identical to persistence.

On the probabilistic interpretation of classical thermodynamics, equilibrium is *nonorder*. To repeat: because there are no constrained interdependencies, there is no organization, no structures, and no configurations. What is usually described as a process that goes from order to disorder is in fact only the progressive relaxation of an existing gradient whose source is unexplained (Barbour 2020)—whether of the Big Bang or a closed box partitioned into two reservoirs. Left to itself, cosmic expansion diffuses until it reaches uniformity, the total absence of constraints. Left to itself—that is, without energetic input that keeps the two reservoirs separate—the gradient dissipates until the molecules of gas are uniformly distributed throughout the space. This form of dissipation is significantly unlike the constrained, order-generating process of nonequilibrium dissipative structures. The two processes are different because according to standard physics micro-events are independent of one another and there are no relations—and therefore no relations of continuity, covariance, or dependence, much less constraint—between them.

Other than the stasis of thermal equilibrium, then, do the concepts of persistence and endurance have any meaning in classical thermodynamics? To return to the dice example, since each throw is independent of the others, a long run of snake eyes is not an ontic entity that persists as itself. "A run of snake eyes" refers only to an observable statistical fluke. If persistence is defined as continuity of the same sort of thing despite changes in composition, is there any provision in classical thermodynamics that accounts for the persistence of coherent entities—entities that in some sense endure despite changes?

The second law of classical thermodynamics is blind to "configurations," that is, to real patterned energy flows (Bejan 2020). Energy does not organize into persistent configurations because there are no contextually constrained relations between states. Because no context-dependent constraints make flows dependent on one another and weave them into internally coherent (because interdependent) dynamics, any such patterns that seem to arise and last are coincidental sequences of microevents. A run of snake eyes. They are not ontologically informed. It is not just that there are no persistent macrostates as such in classical thermodynamics; there cannot be because each momentary microstate is independent of and uncorrelated from the rest.

The second law is therefore silent about the generation of stable but nonequilibrium configurations of interdependent streams of energy flow, their persistence, or their evolution. The constructal law (Bejan and Lorente 2008; Bejan 2020) is meant to extend the second law by addressing precisely this issue, the formation, persistence, and evolution of designs or patterns of energy that flow, in a discernible direction, versus time. The new configurations satisfy the second law because patterns—constrained order—facilitate energy flow.

This book has proposed that operating against a backdrop of context-independent constraints like thermal gradients, context-*dependent* constraints are responsible for the generation of such patterns. Constraints naturally and irreversibly intertwine and organize independent and separate particles and events into coarse-grained types or patterns of energy, matter, and information flow with emergent properties. Contextually constrained relations and interactions set against the background established by context-independent constraints generate metastable and persistent configurations—real macrostates because of the interdependencies they weave among their components. In irreversible phase changes paid for by an overall increase in entropy, constraints interleave streams of matter, energy, and information flow such that overarching constraint regimes with qualitatively distinct properties arise. The coherence of those interlocking interdependencies embodies and enacts newly created information. That information manifests as the emergent properties of the constraint regimes.

In open systems far from equilibrium, that is, ordered configurations emerge out of nonorder and retard thermalization in virtue of the interplay of constraints. From the perspective presented in this book, configurational patterns—their emergence, persistence, and evolution—are the products of context-dependent constraints, operating against the backdrop of context-independent constraints. As a result, identifiable, constrained, multiply realizable interdependencies can endure despite changes. They last as constitutive constraint regimes with emergent properties.

Persistence under nonequilibrium conditions is implemented (top down) by the governing powers of constitutive constraints that confine token realizations to within the basin of attraction of an overarching coordination dynamic—the attractor generated by constraints. As we saw in the illustration of epigenetic landscapes, since attractors are multiply realizable and generated in response to context-dependent constraints, they are not rigid monoliths; over time, their realized tokens explore the adaptive space of the attractor's constraint regime. Coordination dynamics generated by constraints are true examples of the formation of macroscopic order out of nonorder.

This interpretive framework is therefore also about macro–micro relations. The difference with the classical view, however, is that the latter's statistical interpretation does not address constraints; it formulates no principle whereby microevents become interdependent with each other and real macrostates can form and persist as coherent entities, then adapt, and ultimately evolve. In contrast, the perspective presented here considers multiply realizable interdependencies to be real and irreversibly generated in open far from equilibrium conditions by intertwined context-independent and context-dependent constraints. The explanation of constraint closure presented in chapter 7 accounts for these mereological dynamics.

Interdependencies persist because, as contextually constrained coordination dynamics, their constraint regimes coarse-grain nature into types of entities whose governing constraints are degenerate: they can be realized by more than one microstate. Functions such as homeostasis and metabolism are examples of coarse-grained or smoothed-over constraint regimes. So are organisms and lineages; they can be realized by more than one microstate or configuration even as their unity of type persists. Mereological constraints therefore account for why and how organisms last longer than their cells, and lineages last longer than their specimens. Nature generates such multiply realizable types because they have more paths to persistence—to metastability over time—than do the individual tokens or components that realize those types. Multiple realizability therefore also supports evolvability, the capacity to evolve.

The cosmos, we can conjecture, tends to coherence-making by constraints for the same reason that trajectories in the probabilistic interpretation of classical thermodynamics go from hotter to colder: energy flows more easily when constraints align individual energy streams into more coarse-grained dynamics through which to flow. Multiply realizable constrained ontic patterns also underpin the freedom to explore newly formed relational states (Bejan 2020). In short: interactional types are real, not mere observables. Their systemwide properties persist in reality because they are metastable; they are metastable because they embody contextually constrained interdependencies. Such mereological interdependencies, formed by contextual constraints, are degenerate.

From the perspective presented in this book, the capacity for persistence is therefore the outcome of naturally occurring constraints—not, as Boltzmann-inspired thermodynamics would have it, the absence of arbitrarily and exogenously initialized constraints. Local pockets of constrained and enduring inhomogeneities in nonequilibrium are neither improbable and unexplained initial conditions nor observed coincidences. Interactional types or kinds of entities are real products of the organizing capacity of constrained energy flow. Types form and last longer—as real, patterned configurations of energy flow—than either their token realizations or the components and stages that make up those patterns (Nicolis and Prigogine 1977).

Interactional types as conceived here are not universal or eternal. They are induced by context-dependent constraints and realize self-reinforcing, persistent forms that are multiply realized as distinct path-dependent histories and trajectories. By originating from a combination of context-independent and context-dependent constraints, coordination dynamics precede the origin of Life. Form is not synonymous with shape. Forms are context-dependent constraint regimes with emergent properties and powers. They expand dramatically with chemistry and explode with the emergence of dissipative structures, living things, and biological lineages. On this view, conscious beings and the Anthropocene they have created are just the most recent iteration of this dynamic process.

The Principle of Persistence

Analogous to Bejan's proposals, the principle of persistence also aims to extend the second law, from "survival of the thermally stable" to "Nature seeks persistent forms" (Pross 2012; Pross and Pascal 2013; Pascal and Pross 2015). Its authors claim that since persistence (stability in time, rather than the stasis of heat death) is the more general concept, it can embrace the unchangingness of thermal equilibrium as well as temporary and local pockets of form, those improbable macrostates in dynamic nonequilibrium that are realized in physicochemical and biological dissipative structures. The principle of persistence's own arrow of time goes from less persistent and less probable macrostates to more persistent and therefore more probable ones, with thermal equilibrium winning out in the end.

Pascal and Pross recognize that the concept of stability is utilized in two quite different senses, one that refers to heat death, the other to stability in time. As noted, in classical thermodynamics, persistence, strictly speaking, refers only to the absence of constraints—heat death. There is no form or principle of identity at thermal equilibrium. There is no-thing that persists as itself.

In contrast, persistence in open and nonequilibrium dynamical systems consists of self-reinforcing and self-maintaining dynamics that, thanks to constraints, hold fast despite changes at the microscale. Persistence in nonequilibrium open systems is realized as *dynamic kinetic stability* (DKS), not thermal equilibrium (Pascal and Pross 2015; Pross and Pascal 2013). According to the two authors, the source of DKS's persistence is exponential growth such as the growth driven by autocatalysis.

What, exactly, grows exponentially? The principle of persistence considers stability in time to be the ability of certain replicating entities, "to make copies of themselves at a rate that results in a non-equilibrium steady state population of replicating entities being maintained over time" (Pascal and Pross 2015, 16162). In living things, this translates to the ability to reproduce at a rate that outruns entropy or outcompetes other species (Pascal and Pross 2015), if only locally and temporarily. Pascal and Pross, correctly, include nonliving things such as physicochemical dissipative structures in the category of entities that can outrun entropy. They highlight the exponential growth of autocatalysis as the driver of DKS.

The authors also recognize that it is the "*population* of replicators that is stable/persistent rather than the individual replicators that make up the population at any given moment" (Pascal and Pross 2015, 16162; emphasis added). As has been noted above, lineages and populations persist relative to individual organisms, and organisms persist vis à vis individual cells. Mutatis mutandis, this is true of all open nonequilibrium dynamical systems that self-organize in response to constraints. They are multiply realizable, and their token realizations can differ at the microscale despite the continued persistence of the self-organized and coherent whole.

More precisely, then, it is the metastability of multiply realizable and coherent constraint regimes that outruns entropy and persists, at various scales. This is what Pascal and Pross aim to capture with the term *dynamic kinetic stability* (DKS). From the perspective of this book, DKS represents the persistence of coherence, that is, of multiply realizable interdependencies brought about by intertwined enabling constraints and preserved by governing constraints under nonequilibrium conditions. Such continued coherence over time subtends DKS, which is fundamentally the product of enabling and governing capabilities of constraints.

The hypothesis proposed in this book, then, has been that individual physicochemical dissipative structures and biological systems alike temporarily delay entropy by holding fast unity of type—that is, by preserving multiply realizable interlocking interdependencies over time—despite turnover of individual reactants and products, individual catalysts and elements, or individual organisms and cells. Bénard cells persist despite turnover in the individual water molecules; BZ reactions persist despite replacement of individual reactants. This is also so in the case of organisms, of course: the removal and replacement of individual organs, the births and deaths of individual cells, and so on are an integral part of their multiply realizable dynamic. Cells, too, replicate cell type despite mitotic division, replacement of organelles, and so forth. And metabolism preserves homeostatic stability despite changes in glucose utilization, and so on. Throughout, the coherence—the constraint regime—persists.

If the point made in the previous section is correct, however, basing persistence on exponential replication and reproduction and thereby outrunning entropy production begs the question by presupposing the capacity to replicate and reproduce unity of type tout court or, in Pascal and Pross's terminology, form. As noted, form is not identical with shape. As described at the beginning of the book, Platonic forms were historically postulated to account for type identity—for the capacity to persist as the same sort of thing throughout differences and modifications. Forms and Aristotelian and Cartesian substances were postulated as explanations of why spatiotemporally distinct tokens are alike despite differences. Absent Platonic forms, Aristotelian and Cartesian substances, or natural kinds, the principle of persistence must explain the generation and persistence of multiply realizable coherent dynamics, be they realized by autocatalytic replicators or living things.[2]

That is, the principle of persistence must account for DKS's emergence and metastability despite inevitable perturbations and fluctuations. Otherwise, the principle presupposes the persistence it is trying to establish. All positive feedback including autocatalysis can spin out of control. So how do autocatalytic reactions or loops of feedback processes become stabilized into DKS that persists in nonequilibrium? Coherent, self-reinforcing, and self-stabilizing dynamics that preserve unity of type despite remaining in nonequilibrium cannot be presupposed in the explanation by stating that the puzzle of life lies in the types and nature of persistent systems such as some chemical systems present. On pain of petitio principii, it is precisely those types and nature that call for an explanation.

There are no types as such at thermal equilibrium, where, as noted earlier, stability and persistence refer to the absence of constraints. Interactional types as described in this book can hold together and persist in open conditions far from equilibrium only as products of constraint, not their absence. Stating that "even replicative change that appears prebiotically . . . manifests a logical and irreversible drive towards greater persistence" (Pross and Pascal 2015, 16163) begs the question of whence the original capacity for replicative change to make more of the same type of entity, whether of a physicochemical dissipative structures, predator–prey cycles, or species of organisms, all without spinning out of control. As discussed earlier, replication and reproduction are more-makers precisely because they are not cosmic runs of 6s: each replicate's traits are not independent of those of the parent or the previous step. Subsequent steps and generations are conditioned upon earlier ones. It is only because iterations of complex systems are context-dependent and path-dependent that each turn of the exponential spiral makes more of an already coherent structure. This is so for dissipative structures generally, both abiotic and biological.

Exponential replication, in short, presupposes context-dependent constraints.

Whether each microevent is independent of or interdependent with a previous step, both in time and place, marks a significant difference between classical thermodynamics and the coherence-making by constraint inspired by complexity theory. Context-dependent constraints are absent in the first and very much present in the second. The source of endurance of complex systems is the generation and maintenance of form—that is, of coherent interdependencies held together by regimes of constraint. If persistence presupposes the capacity to preserve a coherent structure or dynamic that endures despite changes, proposing to expand the second law by postulating that the evolution of forms displays a direction from less to more persistent begs the question with respect to the generation and coherence of such persistent patterns of dynamic equilibrium—especially since actual realizations of macrostates are not independent of each other.

Pascal and Pross (2015) note "the stability of such systems depends not just on the system but on factors outside the system" (16163). This book has proposed that the role of contextual constraints in generating DKS in the first place is that factor. BZ reactions and tornadoes persist in nonequilibrium conditions as coherent constraint regimes. Without the role of context-dependent constraints in producing dynamic equilibrium, DKS cannot be presupposed in explanations of persistence and form (interactional type, as it is labeled here). Doing so begs the question.

To summarize: exactly what persists at thermal equilibrium and as dynamic kinetic metastability are quite different: contextual constraints underpin dynamic kinetic stability under nonequilibrium conditions; and there are no constraints at all at thermal equilibrium. We can conclude that the principle of persistence begs the question and commits the fallacy of ambiguity along the way; *persistence* (as the authors use the term) spans both thermal stasis and dynamic stability far from equilibrium only if the term is used ambiguously.

Selection by Persistence

Advocates of the principle of selection by persistence present a different understanding of the relation between stability and persistence. This hypothesis interprets persistence as the outcome of a continuous process of selection based on stability. Whence the criteria of selection on this view?

The *modern synthesis*, which merges Darwinian evolution with Mendelian genetics, maintains that random DNA mutations are the sole source of variation from which selection then culls the reproductively advantaged. Over time, the fittest specimens come to predominate. On this view, phenotypes are the product only of accidental genetic mutations.

Since there are "no competing reproducing planets for natural selection to choose between" (Holmes 2019, 34), the modern synthesis breaks down at the level of Earth as a whole. What, then, accounts for the planet's remarkable capacity for long-term self-regulation, a form of self-stabilization[3]—despite, as mentioned earlier, the appearance of photosynthesis, which dramatically changed the atmosphere? Earth's relatively constant temperature range and overall levels of oxygen, carbon, nitrogen, and phosphorus have remained relatively steady over millions of years. Recent discoveries that Earth's geologic activity appears to follow a 27.5-million-year cycle could be considered additional evidence that lasting geological patterns might not be statistical flukes; they hint at the possibility of constrained and evolvable dynamics (Rampino et al. 2021).

Evidence that Earth has managed to persist far longer than living things capable of adapting to their environment have been in existence suggests, in other words, that evolution by selection of the reproductively fit might be only one of several selection mechanisms at work in the cosmos. In contrast to the usual emphasis on random mutations, proponents of selection by persistence maintain that rethinking selection more generally in terms of persistence over time can include abiotic processes as well as living things under one principle.

Once again, the idea of selection implies a macrostate–microstate distinction; it implies a variety of specimens from which the embedding context, a real and constrained macrostate, culls. Previous chapters argued that interactional types span multiply realizable coordination dynamics. In consequence, culling is not a random lopping off. Reminiscent of the craft of tailoring, natural selection is a process of fitting together or rendering compatible an entity and its embedding context. This fitting together is carried out in back and forth cascades of feedforward and feedback loops, from the dynamic's governing constraints (in virtue of which its constitutive constraints remain coherent) to those of the actual token.

Earth's rocky shell, its abiotic atmosphere, hydrosphere and cryosphere, are fitted together with living things to constitute the biosphere, a constrained and integrative, multiscale and multidimensional set of interdependencies. Interfaces between the various levels of organization and dimensions of this complex ecosphere are the active sites where the fitting together process takes place; in the case of the planet, the biosphere serves as the interface between Earth's geological strata and outer space. It actively selects and realizes actual tokens through cascades of top-down negative feedback loops, which change the probability distribution of individual events. Phrased differently, order parameters of the biosphere (be they the dynamic equilibrium of Earth's temperature range, the relative reproductive fitness of living things, or even the distintegrating impact of human actions on the biosphere itself) describe the multiple constraints that the biosphere as interface must simultaneously satisfy. Only those potential tokens that satisfy that constraint regime are selected for persistence.

It is in this manner that constitutive and governing constraints of complex dynamic systems simultaneously stabilize and enhance metastability. They allow autocatalysis's exponential growth to outrun entropy without spinning out of control. As noted, however, the stability of complex systems is not formless, unconstrained, and static thermal equilibrium. Quite the contrary, it is a constantly varying metastability that thanks to a constitutive constraint regime persists as a coherent dynamic in nonequilibrium.

As just discussed, in classical thermodynamics, stability commonly refers to a system's lowest energy state. In contrast, Earth's overall dynamics have persisted for eons as a pocket of constrained metastability away from equilibrium with surrounding deep space. The validity of the hypothesis of selection by persistence therefore turns on a particular understanding of stability.

Its advocates (Lenton and Latour 2018) recognize that the planet's atmospheric conditions are metastable despite being in nonequilibrium with space. Lenton proposes the following. Suppose Earth accidentally arrived at[4] a configuration of elementary constituents that is particularly stable. Such a pattern would not only tend to persist and become self-reinforcing; it would also supersede destabilizing perturbations because stabilized states generate more stability. They do so because, whether physical, chemical, or biological, stable systems as such "not only persist; they get better at persisting over time" (Lenton, quoted in Holmes 2019, 35) by sequentially selecting among degrees of stability. Insofar as stable states "sequentially select" states that are even more stable (see also Henrich 2016), the idea of selection by persistence therefore proposes a form of "sequential selection."

The hypothesis thus postulates that the cosmos must have accidentally reached a state that does not easily and reversibly wash out despite diversely realized configurations or undergoing regular perturbations or fluctuations. The hypothesis also proposes that particularly stable macrostates beget and select even more stable states.[5] Considering that the second law is "blind to configurations" (Bejan 2020) and does not countenance relations among microprocesses, however, how can primordial and improbable but real states not naturally wash out—absent constraints? A primordial and improbable state that holds together either in a principled way, or due to an internal logic, is not contemplated in classical thermodynamics.

In response to critics who question the relative selective advantage of stability over regular and entropic destabilizing forces, proponents of the Stability as Persistence thesis maintain that any stability that persists must be *robust*,[6] that is, able to withstand fluctuations and insults without decohering. What property might delay thermalization and thereby turn the state robust? Without using the terminology of degeneracy or multiple realizability, much less constraint, Lenton notes that certain features are required for robustness.

> Robust systems have some degree of *redundancy* so the loss of any particular component (the extinction of a species, say) doesn't critically compromise the whole. Second, they have *diversity*, which increases the odds that at least some species will be able to cope with unexpected changes. Third they have *modularity* so that a failure of part of the system doesn't bring down the whole thing. (Lenton, quoted in Holmes 2019, 36; emphasis added)

By emphasizing redundancy, diversity, and modularity, the stability as persistence hypothesis implicitly appeals to the notion of macrostates. In light of the previous discussion, this requirement immediately prompts the question, "What mechanism generates real macrostates that are redundant, diverse, and modular?" Without an origin story for these three features, this hypothesis, too, begs the question.

In contrast, the central thesis of this book, that constraints are the agents of coherence-making, offers just such an origin story.

It is evident, then, that just as the advocates of the principle of persistence used the term stability to mean DKS, advocates of selection by persistence use the term stability to mean *metastability*. The robustness required for persistence is a naturally emergent metric of diversely realizable constrained interdependencies. Such variety is not measured in numbers of identical tokens of the same type; it requires interdependencies among interwoven but distinct constraint regimes (Brillouin 1962; Collier 2003b; Collier and Hooker 1999). Integration into a general type of entity in response to constraints lowers internal entropy production and realizes metastability. The more types and subtypes, the more paths to realization. The more deeply multiply realizable and degenerate, that is, the more metastable and robust.

Modularity, the third prerequisite of robustness identified by advocates of the selection by persistence hypothesis, is a common feature of coherence-making by constraints, especially in hierarchy formation. It contributes to a hierarchy's capacity to evolve. Chapter 12 will return to modularity's contribution to coherence-making and hierarchy formation.

10

Sedimentation and Entrenchment

Constraints not only facilitate the emergence and persistence of coherence; constraints can also sediment and even entrench, thereby strengthening the metastability that keeps established, coherent structures from going to thermal equilibrium. In contrast to scaffolding, which commonly aids in the successful completion of a dynamic process, sedimentation and entrenchment preserve existing equilibria in the face of perturbations, actual or potential. *Sedimented and entrenched constraints* can be defined as constraints that are more difficult than usual to relax or otherwise modify. *Difficult* here means it takes more energy to become dislodged than to persist. It takes more energy to get over the hillock of the attractor in which a system is embedded and into a neighboring one.

Sedimented and entrenched constraints therefore strengthen existing governing constraints by making their attractors in possibility space deeper and more pervasive. In living things, sedimented and entrenched constraints reinforce habits and significantly increase the likelihood that only a smaller subset of possible behaviors or phenotypic traits becomes actual (Egbert and Barandiaran 2014). Sedimentation and entrenchment are therefore important add-on, context-independent constraints that reinforce default governing constraints and add extra oomph against thermalization. The long-term price of doing so, however, can be less flexibility and a lowered responsiveness to context.

Sedimentation, Memory, Records, and Registers

Sedimentation confines realized diversity to a subset of empirical possibilities. It can be the product of add-on constraints such as repetition discussed earlier. The concept of *habitus* described by French sociologist Pierre Bourdieu (1977, 1986) suggests sedimented constraints in the tacit and often unconscious but very ingrained skills and predispositions of social class. They can be changed, but it takes added effort and work to do so.

The trajectories of Newtonian mechanics are Markovian. They are path independent; in their relentless march to equilibrium, they forget where they came from. In contrast, context-dependent self-organizing complex systems are path dependent. Because constraints generate path-dependent dynamics, interdependencies woven together by contextual constraints are also forms of memory. The structures and dynamics of coherent interdependencies thus embody records of their context and their history. As traces of their history become integrated into their present constraint regime and continue to influence it, complex systems remember (Rovelli 2018).

In short, sedimentation is a form of memory. Path dependence is a temporal constraint that enables complex forms of memory such as records and registers to arise. Traces of factual information from the past persist in those records; in turn, those that continue to affect the present are sedimented constraints.

Memories can thus become incorporated into governing constraints on possibility space. By reinforcing an existing constraint regime, memories contribute to the persistence of existing dynamics. As records of real constraints, memories also weight and bias future events—even as they are themselves updated in the process. Scars left by volcanoes in Iceland and moraines in Lake Louise, Canada, and Zermatt, Switzerland carry past conditions into the present. We now know that some of these geologic traces continue to produce significant effects on today's climate. They are among the planet's memories; they also carry warnings about fast approaching thresholds of instability where adaptive space might run out.

Ordinarily, memories stabilize constitutive constraints but they can also prevent them from reconfiguring in light of new information. As an example, epigenetic processes such as higher neuronal methylation levels preserve and sediment information of trauma into future generations (Stenz et al. 2018). They harden the constraints of existing possibility space. As noted earlier, obsessive-compulsive behavior betrays a constraint regime that lacks that capacity to be modified. Post-traumatic stress disorder reflects a memory that has become a sedimented governing constraint that is difficult to modify to reflect current conditions.

Bound information in the human genome can be considered the sedimented record and memory of phenotypes that were successful in earlier contexts. Reports that today's human genomes retain traces of our Neanderthal and Denisovan ancestors should come as no surprise. Relatively robust memories, traces, records, and registers can become context-independent constraints that keep conditions away from equilibrium; in doing so, they delay a system's thermalization and preserve it as itself for a longer period before it ultimately succumbs to the second law. In situations where multiple constraints must be simultaneously satisfied, recessive genes that resurface due to interbreeding are likewise evidence of latent constraints. So-called junk DNA might similarly represent a reservoir of latent constraints that surface when other, less entrenched constraints are relaxed.

Organizations and societies share analogous dynamics with prebiotic ones. Customs, mores, values, traditions, institutional memories, and practices are sedimentations of bound information registered by earlier constraints, both context independent and context dependent. Woods (2016, 2018) suggests that hewing to established practices is a successful strategy for an organization under normal conditions. Under stressful and threshold conditions, on the other hand, organizations must position themselves at the edge of adaptive space to more easily spot new gradients or enabling constraints that might be tapped more quickly in the service of renewal and survival.

In sum, as is the case with all ratcheting mechanisms, memory traces and sedimentation are an evolved way of sustaining a coherent structure's interdependencies. Path dependence drags memory into today's constraint regime. Doing so delays dissipation; it preserves a system's constitutive constraints. The past continues to indirectly influence—but under normal conditions and thanks to the multiple realizability of top-down control, not necessarily determine—the present.

Entrenchment

As default governing constraints extend their scope and reach through self-reinforcing mechanisms like cooptation, interactional types held fast by sedimented constraints can become even more metastable in space and time. Red tape seems to follow this pattern. Its tentacles have an inexorable tendency to self-reinforce and spread; as a result, its stipulations become ever more encompassing and last even longer. Red tape entrenches.

Entrenched constraints are even more fixed than sedimented ones. In the service of preserving stability, entrenched constraints reinforce metastability by smoothing over short-term fluctuations and perturbations. By protecting existing systems from short-term crises, entrenchment prolongs their persistence. As with the term scaffold, entrenchment and its cognates can refer not only to the implementation of constraints that entrench processes but also to entrenched phenomena themselves. In the social sciences, entrenchment can also refer to the intention that produces the entrenching artifact, the entrenching constraint, and even the entrenched artifact itself. As an example, the U.S. Constitution serves as an entrenched constraint in this double sense: it was intended to be difficult to alter while simultaneously including provisions through which to do so.

If sufficiently rigid and inflexible, entrenched constraints can freeze conditions altogether. In doing so, they will mimic context-independent constraints by hardening the coordinates, constants, and attractors of the possibility landscape within which local context-dependent constraints operate. Some temporal constraints, in contrast, can promote flexibility by becoming entrenched differentially; they might be in effect only upstream, for example, at the start of long-term projects. Wimsatt (2001, 2007) calls this sort of differential entrenchment *generative entrenchment.*

Why would constraints become entrenched at all? The following section draws heavily on superb and decades-long work on entrenchment (Bickhard 1992; Wimsatt, 2001, 2007; Wimsatt in Caporael, Griesemer, and Wimsatt 2014; Barber 2016).

Why does entrenchment happen naturally? Earlier chapters of this book suggested that improved metastability favors persistence.

The insight motivating the Selection as Persistence hypothesis was that, even in the abiotic domain, governing constraints like those at work in the earth's atmosphere and biosphere became entrenched in virtue of their contribution to dynamic stability; their interdependence entrenched into context-independent constants that stabilized the earth's constraint regime away from equiprobability. Entrenched constraints might have started out as context-dependent constraints; their strong contribution to metastability subsequently caused them to entrench, at which point they became context-independent constraints of an expanded dynamic. This book has maintained that this process is in evidence, mutatis mutandis, in atoms and molecules, physicochemical and biochemical reactions, and even in economies and ecosystems.

Cosmic constraints of the universe's fundamental parameters are prime examples of entrenched constraints. Symmetry and conservation principles, to be discussed in chapter 14, are two fundamental and entrenched context-independent constraints that preserve the universe's basic inhomogeneities. Others mentioned earlier include the gradients of cosmic expansion and gravity, the many forms of polarities, the stability of the atom's nucleus and its electron orbits, and others like the Pauli exclusion principle. These keep the cosmos together by holding universally. As a result, even prior to the origin of life, geological and atmospheric dynamics on Earth remained within a sufficiently steady range of values to persist metastably, neither imploding nor dissipating.

In the social sciences, arguments in favor of deliberate entrenchment often explicitly claim that the default stability of existing human conditions and institutions is precarious and in need of reinforcement. For example, family and social structures might need stabilization to ensure the community's emotional resilience and mental health. If a system's governing constraints are not strong enough, decoherence and disintegration threaten; its viability as that entity might not hold. Hence the appeal of entrenchment.

Deliberate entrenchment might begin with *damping*, which even in the prebiotic realm limits natural fluctuations and prevents runaway acceleration, always a possibility given the role of positive feedback loops in generating coordination dynamics. Negative feedback damps fluctuations; it diminishes the potential for runaway acceleration by reducing the amplitude of oscillations and the speed of feedback loops. Think shock absorbers, both economic and automotive. Damping is a constraint that prevents systems from getting too close to a threshold of instability. In situations where systems might be at risk if they relied only their default rules of change, reinforcing stability through the added constraint of damping might make all the difference in keeping them whole.

As we saw, in biochemistry buffering properties of chemical solutions allow some enzymes to continue to function. In human populations, Old Order Amish practices might have decohered and disintegrated had the group mixed with the general population. Entrenching religious and cultural constraints through the practices of shunning and isolating the population from interactions with the general culture might have been deemed necessary to the group's survival. All of these are damping constraints.

The long-term drawback of damped, buffered, and isolated systems is that these might remain viable only while damped, buffered, and isolated. Their long-term persistence is purchased at the price of staying same, same, same in the near term. Rigid varieties of entrenchment therefore usually provide only short-term solutions; if those added constraints were relaxed, disintegration might quickly follow.

* * *

The value of entrenching constraints deliberately to preserve a system's coherence is another matter. Arguments over the value of entrenchment itself are often based on the presupposition that the system in need of reinforcement is already valuable. In the social sciences, however, arguments for deliberate entrenchment as a general principle are sometimes premised, not on entrenchment's preservation of valuable emergent properties, but on the idea that stability as such is valuable and desirable. Retaining the U.S. Electoral College is often defended on those grounds.

Arguments defending stability as such can rationalize and even conceal their proponents' real goals. Economic and social inequalities everywhere tend to rely on constraints that are self-reinforcing: constraint regimes of existing economic and social spaces self-reinforce through entrenched constraints that increase the likelihood that the existing dynamic will endure. Restrictions to voting access are another pernicious contemporary example of constraints that entrench inequalities. Aside from the obvious ethical objections to such practices, an often-unrecognized consequence of extreme entrenchment is that, however much it might preserve current stability, it tends to inflexibility. The system becomes unresponsive to its inevitably changing context, its needs, and its long-term interests. In the extreme, freezing constraints in place makes interdependencies fragile and brittle by restricting possibility space to one token realization. Crystals are orderly structures, but they are brittle and cannot adapt or evolve. Here too, implications for public policy are evident.

The golden mean between stability and locking up therefore requires a variety of constraint that simultaneously preserves coherence and controls for flexible metastability.

Varieties of Entrenchment

Wimsatt (2001, 2007, and in Caporael, Griesemer, and Wimsatt 2010) and Barber (2016) describe varieties of deliberate entrenchment to preserve social coherence. These varieties differ depending on whether entrenchment is conditioned on (1) the length of time that the entrenchment is constrained to last, (2) the units covered by the entrenchment rule, and (3) the form or manner of the entrenchment rule.

TIME. The length of the period during which entrenchment is active and the units to which its constraints apply describe the entrenchment's reach. A shorter time constraint allows greater turnover. In contrast, the reach or scope of generation-skipping inheritance trusts can govern beneficiaries for a longer period than ordinary Last Wills and Testaments.

Time entrenchment preserves the status quo. Requiring that a given process spend more time in a certain state, slowing down modifications in certain circumstances by adding wait times before change can begin, or even introducing a waiting period between the submission of a proposal and a decision, or between a decision and the start of its implementation, are all examples of temporal, entrenching constraints. As examples, in recent decades, jurisdictions in the United States removed provisions in divorce laws that required a yearlong delay between a petition of divorce and the actual granting of the decree. Before the Dobbs decision that overturned Roe v. Wade, several states introduced constraints that delayed access to abortion by requiring a consultation period, parental permission, and sonograms. The U.S. Senate's filibuster rule was likewise designed to delay a vote on a resolution or a bill. Such changes in temporal constraints can have significant consequences.

UNIT. Entrenchment provisions can also increase or reduce the numbers of units governed by the entrenchment. The scope of the U.S. Navy Secretary's authority is broader than that of a battleship commander. Orders from the former apply to many more individuals than the latter. The U.S. Navy Secretary's authority also has greater depth: it reaches more layers of organization (fleets, battle groups, task forces, etc.) than a platoon leader's authority. Layers of organization will be discussed in the chapter on hierarchy formation.

Taken to extremes, context dependence has the potential to overfit, which renders systems brittle; they become extraordinarily fit, but in only in one context, for one unit. Entrenchment, in other words, tends to render systems unresponsive to changes in context. Overarching and overlapping regimes of enabling and governing constraints can balance entrenchment and context-dependence by applying to different units.

In the political arena, entrenchment rules can protect entire regions and institutions (as units) from encroachment. The U.S. Electoral College just mentioned is an entrenchment measure ostensibly intended to protect low-population regions. It is now reasonable to ask, however, whether the context in which that entrenching rule was meaningful has changed sufficiently since 1787 to make it no longer fit for purpose.

Requiring a supermajority on votes to limit the likelihood of a bill's passage in the U.S. Congress was a different form of entrenchment, one aimed at protecting precarious but existing conditions against non-overwhelming opposition; since then, political polarization has changed the context so much that the provision nowadays primarily reinforces partisanship, that is, the isolation of the two main political units (the Democratic and Republican parties). Whether the U.S. Senate's filibuster and supermajority requirements have outlived their stabilizing benefits is a timely question that highlights the entrenching potential of constraints.

FORM. This variety of entrenchment refers to the way entrenchment rules are made public. Some constraints entrench more transparently or tacitly depending on the form or manner in which they are made known. As a case in point, federal rules and regulations remain conditioned on their formal basis: maritime and administrative regulations cannot supersede the Constitution, for example. The form in which constraints are established can also determine not only the rates and extent of the entrenchment but also the scope of the general principles in terms of which the entrenchment will be governed: "keeping the sea lanes open" describes a narrower scope than "all men are created equal."

The form or manner of publication of social norms often refers to their manner of dissemination. It can be expressed or tacit, for example. Expressly codified laws such as those governing hate speech may or may not last longer than tacit social norms, which are often highly effective mechanisms of entrenchment despite their unspoken character. Only those in the know might understand how tacit constraints work and therefore how to circumvent them to their benefit (Bourdieu 1977, 1986; Di Paolo et al. 2017). Those who lack that knowhow are often easily barred from education and other forms of social capital. Explicit rules, in contrast, have the advantage of transparency (Di Paolo 2017).

Entrenched constraints can crosscut categories. *Crosscutting constraints* can entrench practices and conditions across multiple time scales and spatial regions. Self-imposed forms of entrenchment often serve that purpose. Irrevocable trusts mentioned earlier constrain successors long term: the further reach of their provisions significantly extends the duration of their governing constraints as well as the units subject to those constraints. If the trust also controls the distribution of estate funds, it can simultaneously protect and thwart profligate heirs, for whom those constraints become context-independent constants, for instance.

In some cases, *self-imposed constraints* freeze current conditions by barring future modifications; this can occur when *self-embracing entrenchment* provisions incorporate meta or second-order constraints that control how first-order constraints can be altered. In a process that recalls the strange loops of constraint closure, this form of entrenchment can be set up by defining the default rules of the system recursively to include procedures by which the rules can be changed. Creating entrenchment rules that are revocable or amendable is one form of self-embracing entrenchment that guards against excessive inflexibility and overfitting;

it also protects the capacity to evolve while at the same time safeguarding systemic stability against short-term perturbations. Clauses in the U.S. Constitution spell out the exact procedures by which amendments can be added. Sunset laws and government in the sunshine laws are yet other examples of self-embracing, revocable constraints. Each of these represents different degrees of stringency and contributes to different degrees of flexibility.

Different varieties of entrenchment facilitate different forms and degrees of stability. Evolvability, the capacity to evolve, is incompatible with extreme entrenchment, and freezing undercuts resilience. Some entrenched constraints are intended to fix the minimal degree of stability that ensures viability. In the political realm, the Bill of Rights served that purpose, as did the Interstate Commerce and Freedom of Movement clauses in the 1787 U.S. Constitution (Article 1, Section 8 and Article 4, respectively). These provided entrenching constraints that protected the young nation from disintegration into separate countries.

Minimal stability underpins robustness. But preserving minimal stability alone often will not suffice if long-term persistence in the face of more severe perturbations is not addressed. Additional amendments over and beyond the Bill of Rights became necessary to adapt to changing times. Evolvability, in other words, requires ongoing constraints that actively promote resilience, the capacity to survive by adapting. That is, evolvability requires specific and ongoing constraints that allow the initial constraints to adapt.

In other words, constraints that not only underpin minimal stability but also promote resilience become necessary. This new sort of constraint has been called *generative entrenchment* (Barber 2016; Wimsatt 1986, 2001).

Generative Entrenchment

Complex systems might become paralyzed by the multidimensional, multiscale demands imposed by enabling and governing constraints. Coordination becomes impossible. In contrast to redundancy, which aims to secure fail-safe conditions and prevent failure altogether (an impossibility in a complex world), generative entrenchment instead offers *safe-to-fail* insurance by enabling resilience, the ability to cope with local perturbations, fluctuations, and other insults (Juarrero 1991) through self-modification. This capability requires the golden mean constraint mentioned earlier: constraints that ensure minimal stability of fundamental constraints while simultaneously promoting long-term flexibility. The resilience such a constraint would confer on token and type alike would greatly enhance evolvability, the capacity to evolve.

One way generative entrenchment accomplishes this is through *differentially expressed constraints*, temporal constraints that activate differentially, in sequence, conditional upon timing and context. Generative entrenchment, for example, might impose rigid constraints early on in a long process while supporting more flexible downstream constraints that open new options and facilitate new interdependencies.

Consider the following example. Epigenetic disruptions such as trauma or starvation early in pregnancy often cause miscarriages. Generatively entrenched developmental constraints protect against that scenario by imposing stringent context-independent constraints early in gestation; these ensure that the fundamental structure of the phyla's morphological features (its basic radial or bilaterally symmetrical body plan, for example), is strictly satisfied. Once that *bauplan* is secured, those inflexible constraints relax and others more responsive to current context are enabled. Mutations and epigenetic influences later in developmental processes are often not seriously deleterious; indeed, it can be argued that the flexibility supported downstream during ontogenesis allows phenotypic possibility space to be explored and new structural and behavioral variations to be generated (Wagner 2014). Generative entrenchment thus narrows and even closes off options altogether early in a diachronically constrained dynamic while simultaneously facilitating unanticipated interactions and other enabling constraints to appear and operate downstream. The process makes room for flexibility, but ensures it only happens downstream.

Viewed at short range, generative entrenchment might appear to be a fragile strategy because "any attempt to change [constraints early in the process] leads to the collapse of existing structures that depend on it, so modifications don't survive" (Wimsatt 2007, 134). In the long term, however, differentially expressed constraint regimes that enact generative entrenchment are self-ratcheting mechanisms that promote resilience. Once each stage stabilizes, the system can explore and contextually probe local and current possibility space without running the risk of spinning out of control or falling apart. The activation of more relaxed constraints downstream then allows local and more timely constraints to construct new coherences and niches with emergent properties and powers—as the current context affords. Generative entrenchment recalls Lila Gatlin's (1972) thesis about vertebrates figuring out how to stabilize some constraints while allowing others to evolve.

Differentially implemented and generatively entrenching constraints underpin robustness and dynamic stability (Wimsatt 2007, 134–135). They underpin evolvability, in other words. They promote robustness and resilience through the spatiotemporally differential activation and inhibition of enabling and governing constraints, as the occasion requires. Think cooptation in prairie grassland ecosystems and in the metagenetics of the gut microbiome. Generative entrenchment can accommodate Stuart Kauffman's (2014) "unprestatable" contingencies while staying true to type. It can also account for *exaptations*, which are not the outcome of mutations; they are functional properties that emerge as a newly coherent dynamic generated by contextual constraints operating in a different context, at a different moment in time. Stephen Jay Gould's spandrels of San Marco (Gould and Lewontin 1979) configure the possibility space for a beautiful new artform to emerge even as entrenchment ensures the dome does not collapse.

Generative Entrenchment of Validated Constraints

A stronger and more significant argument for not only the persistence but also the growing spread of generative entrenchment in living things is that its constraints tend to stabilize only those emergent forms of coherence that represent validated evolutionary transitions (Wimsatt 2007). In its capacity as aide-mémoire, generative entrenchment recalls the original reason for entrenchment and reaffirms the value of the entrenched (Wimsatt 2007).

The value of generative entrenchment, therefore, is that only those processes, constraints, and dynamics that have been validated as enabling and sustaining the capacity to evolve become generatively entrenched (Wimsatt 2007; Kirchner and Gerhart 2005). Only constraints that generate and maintain resilience become predominant over time—because only validated forms of coherence are resilient and self-reinforce evolvability.

In short, the capacity to evolve is likely to be conditional in no small measure on generatively entrenching forms of self-constraint.

Stated otherwise, constraints become generatively entrenched only if they support resilience. Resilience confers selective advantage thanks to its capacity to persist as a coherent unit across a wide range of circumstances and to evolve over longer periods of time. The ratcheting character of resilience, however, requires a generative and differentially activated constraint regime. Acting as a constant, such a constraint regime ensures minimal survival upstream while simultaneously relaxing restrictions on realizable possibility space downstream in response to more local and timely context-dependent constraints. Generatively entrenching constraints are therefore responsive to context and support evolvability even as they underpin continuing coherence. Together, the constraints embodied in the genome and epigenome satisfy these requirements.

A central conclusion of this work is that, together with its emergent capacity for top-down *analog control* (to be discussed in next chapter), coherence-making through contextual constraints is a validated evolutionary innovation that became generatively entrenched (recursively) because of its contribution not just to persistence but, actively, to resilience and evolvability. Ensuring minimal stability while simultaneously opening future possibilities conditional on context is among the cosmos's more recently evolved varieties of constrained coherence-making.

11

Many-to-One Transitions, Effective and Analog Control

Phase changes from separate and independent entities to coherent dynamics are sometimes described as discontinuous Set-to-Superset (Grobstein 1973) or Many-to-One transitions (Patten and Auble 1980). As reviewed in earlier chapters, catalysts and feedback loops, recursion, iteration, and various forms of closure are enabling constraints that induce such symmetry breaks. This chapter makes three key points: (1) Coherence-making by constraints brings about Many-to-One transformations. These Ones are the coordination patterns we have called interactional types. (2) Interactional types are emergent dynamics characterized by continuous properties. (3) Because continuous properties take a range of values, they support a new form of "causation": analog processing and output control.

Integration, coordination, and streamlining intertwine separate entities into continuous dynamics. Synchronization, resonance, alignment, entrainment, phase-locking, homeostasis, and other forms of metastable coherent dynamics are the products of enabling constraints that smooth separate and independent flows of energy into continuous configurations dilated in time and space. Coordination dynamics are the outcome of constrained interdependencies and are enacted as streamlined energy flows. From the physical, through the physicochemical, to living things, ecosystems, and social organizations, different constraints, implemented differently, result in different types of continuous relations with qualitatively novel emergent properties and behavior.

In nonscientific contexts, the concept of *parameters* refers to the collective properties of such complex dynamics. These include vorticity, frequency, and different forms of metastability described earlier. Social parameters include social cohesion and economic equality. The Gini coefficient, for example, measures the inequality in a society's income distribution. Phase transitions from many-disparate-to-coherent-Ones can be considered discontinuous transitions to new possibility spaces—new topologies—that are characterized by novel and continuous parameters. Values at one level become variables in the next.

Order parameters best describe such systems. These are relatively low-dimensional variables that capture the significant emergent properties of those new interdependencies. Order parameters measure, for example, the degree of cohesion of a system's interdependencies, a continuous property that is qualitatively distinct from properties of the relata separately. *Order* here refers to the degree of coupling and mutual dependencies among a system's components. Degrees of synchrony and metachrony, for instance, represent different interdependencies and, as a result, different degrees of coherence. As soldiers' marching illustrates, intensity and relative phase correspond to degree of coherence. Partial synchrony reflects incomplete coherence; social polarization describes its almost total absence. Judges of synchronized swimming competitions evaluate performance in reference to such order parameters.

The discussion about self-simplification in chapter 7 proposed that terms like *social capital* and *economic development* are not mere simplifications that discard detail for epistemic convenience. The collective properties, order parameters, and governing constraints of those emergent Ones are qualitatively distinct relational properties that differ from those of the many separate individuals that constitute the interdependencies. Seen through the lens of complexity theory, terms like homeostasis and social capital encode high-dimensional data into intertwined and lower-dimensional constants. They are therefore best understood as "variables of a lower dimensional state space" that encode order parameters of real interdependencies formed by enabling constraints (Hinton and Salakhutdinov 2006. Also see chapter 15).

In living things, Many-to-One transitions are uncontroversial. Signals impinging on living things are calibrated and filtered in terms of order parameters that represent the organism's integrity (its coherence) as well as how well those signals fit its current habitat or niche. Consequently, organisms do not respond in the same way to the same physical input; different physical inputs can elicit the same output. To illustrate, interoceptive as well as sensory signals are adjusted by allostatic mechanisms[1] with reference to parameters pertaining to viability and health, given those circumstances. In consequence, different metabolic and endocrinological processes recalibrate and reset such that they continue to support the organism's order parameters of viability and health, adaptability and evolvability.

Allostatic mechanisms, in other words, are governed by constraints that select, discard, harmonize, and calibrate processes such that they satisfy the multiple constraints set by overarching order parameters, which were in turn generated by enabling constraints. Combined, the multiple constraints of those parameters set the conditions of satisfaction for the token realizations. Moment to moment, circumstance by circumstance, coordination dynamics that constitute homeostasis adjust the probability distribution of the relevant variables with a view to realizing the range of values that represent viability and health. Selection and interpretation describe this multilayered process of multiple constraint satisfaction. Timing, amplitude, activation, suppression, and so on of the various variables are recoded, recalibrated, and reset in real time in reference to the organism's order parameters. Stated otherwise: governing constraints of homeostasis and allostasis tune and select signals with reference to metrics pertaining to the system's overall order parameters, metastability and viability: "The [biological system] has in effect produced a *Many-to-one input-output function* by *reinterpreting* the signals from its environment. That is, the [system] has made a *model* of *effective* (phenomenal) *input* Z_t from potential (physical) input Z'_t, based on its state at time *t*" (Patten and Auble 1980, 161; emphasis added).

For Patten and Auble, Many-to-One transitions in living things involve reinterpreting physical signals in terms of a model. Analogously, Pattee (1972a, 1982) describes effective output as response in terms of an abstraction. As mentioned earlier, the ability of living things to "statistically classify" in this way motivated the new field of *biosemiotics*, of which Pattee was a founding member.

What Patten and Auble call *effective inputs* and Pattee calls an *abstraction* are therefore raw physical signals recoded as variables that matter for those living things at that moment, to paraphrase South African cosmologist and Templeton award winner G. F. R. Ellis (2016, 2021). Transforming physical signals into effective input is a process of contextual selection carried out with reference to a coherent set of interdependencies and affordances. Early English settlers on the American continent, for example, did not perceive large tree trunks as affording house building; only milled wood planks functioned as effective input. Scandinavian settlers, in contrast, readily perceived whole logs as affordances for residential construction. Like habitats, such emergent affordances (Gibson 1975) represent indexical properties that, as effective input, implicate the entire ecology in which signal and subject are embedded.

Order parameters thus describe the interdependencies and governing constraints of the encompassing One (the organism's overarching constraint regime) they characterize. The organism's effective input and output are tuned, not to a signal's physical features alone, but to the organism's constitutive constraints and in the service of their collective properties and order parameters. In other words, organisms register internal and external signals as recoded in terms of their contribution to the emergent properties of the organism's overarching constraint regime. Processing information by recoding signals into self-organized continuous variables in this manner consumes less energy than digitally processing each separate and individual raw physical signal every time. It is also more noise tolerant.

Brain development illustrates this pattern. In infancy, developmental constraints prune previously jumbled neurons into complex and streamlined networks. As depicted in Fig. 4.1, infants learn to walk in response to a combination of constraints, primarily gravity and the child's increased weight. The consequence of such multiple constraint satisfaction is that the infant's jerky flailing becomes coordinated into a toddler's stepping pattern, characterized by new properties and powers. The new dynamic also shows lowered rates of internal entropy production (Thelen and Smith 1994). The emergent coordinated stepping pattern is thus the outcome of a phase transition from separate and uncorrelated jerky arhythmic movements to coordinated and rhythmic walking. Coordinated and smooth walking enact the continuous order parameters of this new dynamic. All in response to constraints, in satisfaction of the second law.

We can conclude that coherence-making generally is the generation of such Many-to-One functions. Fixed discrete relationships at one level become systemwide and continuous variables at the next integrative level (Allen and Starr 1982, 241). Coherence making thus brings into being a novel coherent One that must now be mapped in a relational, parametrized space. These relational properties, parameters, and powers are real and effective; they are not just the sum of internal primary properties. They represent novel relational constraints that matter for the system's metastability and viability, and they exert their influence as constraints, not forceful impacts.

* * *

This work has repeatedly emphasized that Many-to-One phase transitions have precursors in the abiotic realm. Many-to-One transformations generally integrate separate and discrete entities into generic and continuous, streamlined dynamics. Such transitions take place even among organisms in different kingdoms. Lichens are coherent Ones that emerge from the interdependence of algae and fungi. Even in the abiotic realm, streamlining and recalibrating in terms of order parameters results in more energy efficient and noise-tolerant metastability. We have described it in convection cells, laser beams, and BZ reactions.

It is hard to believe that exactly what covalent bonds are, and how they form remains controversial to this day. Molecular orbitals depend on covalent bonds between their constituent atoms. One theory maintains that covalent bonds are constituted as constructive interference patterns between atomic orbitals (Nordholm and Bacskay 2020). Iterated resonance integrates the initially separate reverberations and oscillations of atomic orbitals into expanded patterns of constructive interference, the novel and coherent coordination dynamic we know as molecular orbits. Once entrained into this more encompassing dynamic, the motions of erstwhile separate components align in the service of the new collective dynamic. Rephrased, covalent bonds are coarsened interfaces of a more encompassing, contextually constrained attractor or energetic pathway. Emergent Ones like these, whose continuous properties embody a distinct constraint regime, define a new domain, with new types, properties, and powers. Viewed through this lens, molecular orbitals are streamlined smoothings of the atomic orbitals of which they are composed. The order parameters of molecular orbitals control, top down, the behavior of their component atoms.

Analogous Many-to-One transformations also occur in social and ecological relations. In human organizations, enabling constraints such as bonding, bridging, and linking together of individuals and groups generate mutually constrained and constraining interdependencies that, as emergent Ones, constitute social networks, social norms, and social trust, each a new collective variable with its own set of order parameters. In turn, the degree of coherence among social networks, norms, and trust underpins distinct levels of systemwide social cohesion, economic development, ethnic and racial tension, and social capital. Each of these terms compresses ever more complex interdependencies and order parameters formed by constrained interactions among the lower-level processes; each also describes a higher-level coordination dynamic, with novel and significant properties and powers. In each case, the new contextually constrained and contextually generated configurations reduce rates of internal entropy production while allowing energy to flow more easily overall. Many-to-One transformations everywhere are nature's way of streamlining through coherence-making by constraints, once again in satisfaction of the second law.

Described as *ecological fitting* in the biotic realm, constraints adjust and integrate living things to their current environment—in light of their past, as the occasion requires. They generalize by creating more encompassing constraint regimes. Intertwining and smoothing over disparate energy streams into shared coordination dynamics creates biological habitats and niches as well as molecules, each with emergent continuous properties. In each case, interfaces between entity and context coadjust input and output to the coherence of the extended, metastable dynamics.

The more complex the life cycle, the more overlapping and multidimensional the constraint regime and the more expansive the possibility space, the more complex that emergent One becomes. Social animals like bees and primates process more complex information than simple, nonsocial organisms, but all complex systems, abiotic and biotic alike, continuously generate and enact analogous interdependencies. Our failure to understand mereology is one reason why we have remained ignorant of the complex coherence of mycorrhizae, the symbiotic association between fungi and plants.

"Ecological contexts, for example, become synchronized with development and reproductive activities," states North American parasitologist Daniel Brooks (email communication 2021). He illustrates: *Crithidia* are parasites that ordinarily live in plant tissues. Because the intestines of plant-feeding insects contain plant juices, they represent an overlapping habitat[2] with the original plant-feeding *Crithidia*. Sufficient overlap between the constraint regimes of plant tissue and insect intestines allowed *Crithidia* to jump the gap to plant-feeding insects themselves. *Constraint overlap* enables a novel set of interdependencies—in this case, an expanded *Crithidia–insect* niche or habitat.

In a process reminiscent of causal closure, strange loop-like mereological relations like these constitute and realize complex, multiscale intertwinings: in this case, the parasites circulate between the plants and the plant-sucking insects. Are the plants an extension of the insect gut or does the insect gut represent an extension of the plant (Brooks, personal email communication 2021)?

As we saw earlier, by putting out meristems, grasses effectively coarse code and smooth over criteria for signal integration and coordination into a broader set of constrained interdependencies, with new order parameters. Doing so generates a new Many-to-One function, a new interactional type. An analogous process can be described for chaparral, the oil-laden succulent prone to burning (Allen and Starr 1982). Far from destroying an ecosystem, chaparral brush fires clear out the underbrush. (Note that chaparral does not bring about change as efficient cause: chaparral does not actively start, quench, or block fires.) By providing a path along which energy can flow more easily, chaparral instead facilitates the eruption of small brushfires; these occur more frequently thanks to chaparral's oily composition. Critically, far fewer large and destructive wildfires break out as a result. Eventually, the enabling constraint provided by the presence of chaparral becomes integrated into the constitutive and governing constraint regime of a more metastable One—in this case, an expanded ecosystem with new longer-term generative properties that the earlier ecosystem lacked. The change occurs thanks not to efficient causes, but to effective input–output relations generated by constraints.

It is not too farfetched, I submit, to note in this connection that the possibility space of interdependencies that define parenthood have likewise coarse-grained in recent years, from an interface defined by bloodline-based criteria to one that is caring based—a Many-to-One transformation like the *Crithidia* and chaparral examples. As constitutive and governing constraints of the interfaces between a society and its context become more coarse-grained in response to more encompassing enabling constraints, affect and caregiving became parameters that "matter" as much as blood. A more encompassing set of interlocking constraints and interdependencies between biological parents and other childcare-givers formed. These constitutive constraints underpin new order parameters and distinct metrics that now matter. Epistemic models and social regulations change accordingly in response to these more encompassing and coherent interdependencies.

In sum, Many-to-One transitions in abiotic and biotic domains alike reshape possibility landscapes by extending the scope and reach of coordination dynamics, along with establishing new governing constraints and order parameters—all in response to the workings of enabling, context-dependent constraints against a backdrop of context-independent constraints. By coarsening the gating criteria of the interface between the original grasslands' niche and horses, between *Crithidia*'s original plant-hosted niche and insects, between two atomic orbitals, or even between the relationships between adults and children, novel interdependencies, parameters, variables, and values continuously adjust such that their new and overarching constraints are simultaneously satisfied. A more expansive One—and more metastable domain—coheres. As a new identity that is more robust, multiply realizable, resilient, and persistent colaesces, the possibility landscape becomes more individuated and rugged.

Patten and Auble and Pattee's notion of phenomenal experience as an interpretive model or abstraction can therefore be generalized as follows: models and other abstractions such as order parameters represent constraint regimes that filter, code, and calibrate internal as well as incoming signals in terms of systemwide and continuous relations of dependence. Those constraint architectures not only establish entry points into the existing dynamic; because they are contextually constrained, their order parameters also recalibrate and are recalibrated by and with each new set of signals and outputs. Consequently, the system's overarching constraint regime's architecture and order parameters continuously reconfigure and persist—as the constitutive and governing constraints that matter. Processes covary such that the constraint regime remains locked in to real-world distinctions and therefore persists.

Phase transitions from separate manys to coordination dynamics that are streamlined and smoothed coherent Ones—epistemically or ontologically—are therefore forms of generalization (Dean, personal email, 2021). Contextual constraints in open systems far from equilibrium generate new equivalence classes (Ellis 2016, 2021) with novel powers of top-down control. Public policy, for example, represents a society's recoding of individual events and processes into a coherent constraint regime of equivalence classes that matter to that society. Its constitutive and governing constraints are implemented through interfaces such as codes and rules that make sense of the overarching policy in context: they recode, filter, and translate signals (such as actual rates of immigration) in terms of the culture's existing social and economic networks, its norms, and levels of trust. Translated into laws and regulations, these represent the society's constitutive and governing constraints; they effectively establish the variables that matter to it and that control its actions, top down, such as to maintain and preserve its existing constraint regime.

Effective input to any coherent dynamic therefore occurs when internal and external signals are calibrated and tuned to the order parameters of a set of interdependencies, current or anticipated. Those interdependencies sift, filter, calibrate, and harmonize incoming signals, entities, and processes with reference to their capacity for integration into the existing order parameters; effective inputs and outputs then covary with respect to those order parameters. This Many-to-One recoding occurs in synchronized dynamics, constructive interference patterns, habitats, cultures, and interpretive frameworks. Accordingly, input must be standardized, transduced, and harmonized—recoded—for inclusion and selection in light of the dynamic's interdependencies. Seen through this lens, Many-to-One transformations predate the origin of life. They are the outcomes of constraint that harmonize and streamline energy flow by integrating spatiotemporal processes into context-dependent coordination dynamics characterized by continuous variables. When not only "stability" and "homeostasis," but also "value, ethics, and morals" (Artigiani 2021) become the interdependencies that matter, ensuing behavior is more sensitive to social context. Most important of all, the actions governed by those constitutive constraints will enact the novel emergent properties of those social parameters. Context- and path-dependent systems with emergent properties with the capacity for top-down control show that a reductionist approach will ultimately fail. Patterns of interwoven constraints cut the world along a different set of joints.[3] Actually navigating a reorganized space requires that organisms process the enabling and governing constraints that sculpt the transformed topology's new attractors and specify its transition and selection rules. Epistemically, in other words, finding one's bearings in a newly reconfigured or reconfiguring topological space is inevitably a process of sense-making through narrative (Snowden 2015) and hermeneutics (Juarrero 1999).

* * *

Setting the gating criteria of more expansive (coarsened) interfaces in terms of new codes facilitates Many-to-One transitions. A *code* is a set of instructions or rules (constraints) that translates or transduces signals into a different format. Individual signals on one "side" of an interface are recoded into the format of the integrative dynamics on the "other side." The genetic code, for example, translates molecular structure into biological function. Recent neuroscience experiments described in chapter 15 show how the cortex of macaque monkeys recodes individual neuron responses into continuous waveforms of invariant phase and amplitude. Chemistry, too, is a new code that transduces physical entities into novel interdependencies with continuous properties. Likewise, values, ethics, and morals are evolutionarily recent codes that do the same. They translate and integrate signals in terms of the constraint regimes and order parameters of human social organizations (Artigiani 2021)—which embody the variables that matter to a culture. This book has emphasized physicochemical and straightforwardly thermodynamic processes only to drive home the work's goal of naturalizing metaphysics; the principles are meant to generalize to living things overall and human beings in particular.

Tagging Emergent Properties

Evolving interfaces that operate according to new codes facilitates multimodal and multidimensional integration and provides survival advantage. Many-to-One transitions recode signals as integrated and continuous relations. For social organizations, survival advantage is implemented in terms of emergent systemwide properties. Accordingly, criteria for group selection refer to patterns of behavior that contribute to the group's survival and metastability, given the overall society in which that group operates. This section of the chapter will hypothesize that in the case of sentient beings, tagging the recoded integration as emergent phenomenal properties and then acting in response to those tags considerably improves the organism's effective decision-making and action. The sequence might go as follows: Many-to-One codes facilitate phase transitions that integrate a coordination dynamic with novel properties. Tagging those emergent relational and continuous properties as feels (*qualia*), phenomenal images, and ethical norms would significantly contribute to more effective decision-making and actions for the organism as a member of society. Effective input and output newly coded and tagged in this manner would enhance metastability.

* * *

Many of the brain structures in the limbic system such as the amygdala (anatomical association nodes that integrate information from a variety of sources) translate incoming signals in terms of new codes. What would such codes look like? They begin, arguably, with the code of *valence* (Damasio 2021). The limbic system in general integrates and recodes separate streams of information as valence, as the "positive or negative appraisal" of the effects of integration; henceforth, this evaluative parameter becomes the variable that matters.

I propose that tagging those axiological parameters as qualitative feels (qualia) further enhances effective decision-making and behavioral control. Faster and more energy efficient to process, valence-encoded dynamics tagged as qualitatively distinct sensations would streamline energy flow and also satisfy the second law.

Complex contextual considerations recoded in terms of valence and perceived as feels, for example, can quickly reinterpret and transform an enjoyable friendly tickle to the negative experience of a perceived assault. Integrated information about the sources and meaning of particular experiences, recoded and tagged as qualia, can transmit multimodal information about complex current conditions much faster and more comprehensively than cognitive processing alone. The faint sound of a doorknob turning unexpectedly at 4 a.m. will elicit the gut feeling of danger and awaken us when loud but regular airplane noise overhead does not. The emergent properties and constraint regime of the perception of danger are multiply realizable and continuous; they integrate factual, sensory, and evaluative aspects of an experience. Since continuous properties are noise tolerant, false positives are always possible, but overall, one can hypothesize that the benefits in terms of metastability would outweigh the negatives.

In addition to the amygdala, the brain's anterior cingulate cortex, thalamus, and insula also recode a combination of past records and current stimuli. The resulting continuous properties of that smoothed and streamlined dynamic likewise become tagged as varieties of negative qualia, fear, anger, or pain. In parallel, the *ventral tegmentum* on the floor of the midbrain and the *nucleus accumbens*, a region in the basal forebrain, recode stimuli in terms of positive valence (Damasio 2021) and tag the result phenomenally as, say, pleasurable, flavorful, or fragrant.

Next, top-down control governed directly by qualia and phenomenal properties further facilitates decision-making and action. That is, directly integrating these phenomenal tags with motor activation or inhibition centers further improves decision-making and output control mechanisms for action. In this fashion, putrid odors link to regions of motor control to produce action; the feeling of fear is inextricably intertwined with the propensity for flight behavior. That is, behavior appropriate to a particular valence—fight or flight, avoid or approach—follows quickly and directly from such recoding, integration, and phenomenal tagging, tied to motor control. But only at this moment, under these conditions, given this history.

It is reasonable to hypothesize, therefore, that from Many-to-One transitions to continuous type-level properties there evolved in living things a new form of selection and behavioral control that is implemented with direct reference to qualitative feels and phenomenal awareness. For example, decision-making and action (1) carried out in reference to a continuous parameter, danger; (2) tagged as a qualitative feeling of fear; and (3) coupled with motor processes produce *RUN!* This intertwined skein of constraints improves survival advantage precisely because it speedily conveys information about a variable that matters—namely, a threat to survival. No need to overthink the feel. Just run. In contrast, lack of such phenomenal awareness, as in *anosognosia*, the inability to feel pain, puts its sufferers' lives at risk. Failure to integrate phenomenal awareness and feelings into motor processes would be otiose.

Damasio (2021) notes that interoception and proprioception might have been the earliest phenomenal properties with indexical properties to emerge from multimodal integration and coordination. By tagging integrated information phenomenally, the qualitative proprioceptive feel of "the twisties" quickly enables gymnasts to correct the position of their body in space. This indexically coded "sixth sense," which might even precede evaluation in terms of valence,[4] provides a critical safety measure that prevents bad spills. Proprioception and interoception integrate the material conditions of the subject's body, its location in space, as well as an individual's sense of ownership of it in terms of the organism's actual and continuing metastability (Damasio 2021).

Two of the most influential theories of consciousness today, the global workplace theory (GWT) (Koch 2004; Metzinger 2000) and integrated information theory (IIT) (Tononi 2008, 2012), take integration (of sensory data or Shannon information) as a necessary and/or sufficient condition for the emergence of conscious awareness and phenomenal feels. On the way to global integration, however, less comprehensive coherent structures, transduced by new codes and tagged with phenomenal properties, might have become modularized as network nodes and modules of the overall dynamic. Context-dependent networks, that is, can become modularized into context-independent nodes that stabilize conditions and delay thermalization.

The mammalian brain's early sensory cortices as well as its association cortices can be understood in that light. Having evolved later than the amygdala and the nucleus accumbens mentioned earlier, the *fusiform gyrus*, an anatomical structure in the temporal cortex, also filters discrete visual stimuli and integrates them as faces (a continuous property) and tags them as a distinct three-dimensional and emergent phenomenal image (another continuous property). Coding individuals in terms of gait (yet a third continuous parameter) might be an even more accurate way of recognizing individuals (Adam 2020). This is consistent with the fact that over a person's lifetime, gait is more metastable than facial appearance. Identifying individuals over long periods of time and over a long distance without direct contact suddenly becomes possible thanks to metastable, multiply realizable, coherent dynamics.

Over time, chances of metastability and survival either improve with each evolutionary iteration or the system dissolves. The system either adapts and evolves, or disintegrates. Even the *Mimosa pudica* plant habituates and stops shrinking when it senses a familiar stimulus it has learned to categorize as nonthreatening. This is how anchoring, priming, and cueing occur in human perception and action. Attentional and change blindness errors are direct phenomenal outcomes of emergent coherent dynamics, their governing constraints, and effective input. Subjects are not consciously aware of the gorilla because they were primed—cued—to perceive only effective input, that is, only those visual signals relevant to a task space intentionally and intensionally defined, following the bouncing ball. False negatives do happen. Like affordances, anchoring reconfigures the probability distribution of mental frameworks. In consequence, the behavior satisfies the constraints of the emergent properties that define the task.

Effective output is therefore output continuously updated and controlled by new codes and governing constraints of coherent Ones in which the dynamic system is entrained. Actions performed by coherent Ones enact the bound information in the interlocking constraints that define its dynamic. Implemented as top-down cascades of negative feedback loops, actions enact and preserve those governing constraints. As a result, actions effectively aim in a direction that prolongs that constraint regime's existence—toward heat and moisture in the case of tornadoes and towards carrying out the intention to perform a particular task, in ours. In this manner, teleonomy and teleology get built into constraint regimes thanks to enabling and constitutive constraints.

As the coordinating of mutual interdependencies becomes increasingly complex, we can hypothesize that actions informed by ethical and moral values are performed analogously (Artigiani 2021). Evolving and recoding society's order parameters in terms of symbols and aesthetic and moral values came to characterize human culture. They become the variables that matter. The feeling of moral disgust, for example, might be an evolutionary extension of the repugnance experienced when encountering putrid meat (Aznar 2021; Kurth 2021). Newly evolved normative parameters integrate and filter information with reference to those higher-level variables that matter to a society's constraint regime.[5] Feelings of pride and shame encode much more complex dynamics centered around social information. They identify those who experience those feelings as members of a community. In the process, recoding experiences in terms of value codes and then acting for symbolic, religious, ethical, or patriotic reasons enhances social cohesion and group selection (Artigiani 2021). They better serve as coordination constraints. But only if the agents act for those reasons, that is, only if their actions issue top down from those higher order constraints. Failing to take into account such complex social, religious, patriotic, or tribal sentiments, feelings, and phenomenal experiences would, as noted earlier, get an agent's or a culture's constraint regime seriously wrong.

Analog Control

Since constraint regimes of coordination dynamics realize continuous order parameters, Many-to-One phase transitions suggest that analog decision-making and effective output control might have provided significant innovations for coherence-maintenance and behavior. We turn to the topic of analog control next.

British cybernetician Ross Ashby[6] formulated the Law of Requisite Variety: "Any control mechanism must have at least as many states at its disposal as the thing it is trying to control" (Ashby 1958). In the real world, one-to-one cause–effect relations of strict determinism are fragile and easily disrupted by severe perturbations or fluctuations. They do not generalize. At the other extreme, flexibility alone is not sufficient. With a third of their neurons spread out throughout their body and tentacles, octopi find coordination unwieldy; from an evolutionary perspective, acting not just flexibly but effectively—from an integrated and coordinated space—becomes necessary. Even today, coordination and integration remain the primary obstacles for digital AI systems, including those enhanced with machine learning (Mitchell 2021). In living things, a central nervous system with a brain protected by a skull and capable of supporting context-dependent coordination might have evolved in response to such ecological pressures. A central point of this book has been to emphasize the integrating, stabilizing, and coordinating functions of constitutive constraint regimes.

Analog control mechanisms facilitate such coordination as well. Let us see how.

* * *

The functionality of digital mechanisms like toggle switches, with one crisp output per discrete input, is neither robust nor resilient. Digital systems cannot scale up to satisfy Ashby's concerns. This book has argued (1) that generating multiply realizable pathways while simultaneously ensuring coordination and preservation of unity of type constituted a major transition in evolution for open systems in nonequilibrium. It also argued (2) that constitutive constraints allow top-down control, as constraints, not efficient causes. The next section proposes that (3) processing and controlling output in reference to continuous values supports a wider range of behavior. The notion of analog control will be used to make this point.

Analog artifacts and dynamics implement a Many-to-One logic. For example, by analyzing input as logical variables and not discrete voltage signals, fuzzy logic thermostats take continuous values between 0 and 1. Then, by calibrating their output to those continuous values, analog thermostats do not overshoot their mark and are more responsive than digital thermostats; their functionality is greatly improved. In real time, analog mechanisms are more context and task appropriate than digital mechanisms. To illustrate: integrating two previously independent variables into a continuous order parameter reveals an emergent (because relational) property that really matters to automobile drivers: whether the vehicle's wheels will lock and cause skidding. Analog antilock brake systems (AABS) prevent skids by adjusting braking pressure in reference to values captured as analog order parameter wheel lock, a descriptor of the relation between braking and vehicle speed.

Because their settings and governing constraints refer to continuous, population-level properties, analog-coded top-down control mechanisms can implement meaningful and appropriate output that preserves proper and effective function under a much wider range of conditions. Analog top-down control is therefore critical to sustaining long-term coherent interdependencies. By modulating a deep continuum of realizations, analog-coded devices can satisfy parallel and multiple constraints over a longer period, under a wider set of circumstances.[7] Top-down control from coherent interdependencies to token realizations is executed by analog governing constraints: the order parameters of social organizations and long-term intentions operate as governing constraint regimes encoded in terms of continuous properties such as social cohesion, values, and purpose.

Coordination mechanisms in analog form are well suited to execute complex top-down control. Because the contextual constraints that generate interactional types are not only multiply realizable but also noise-tolerant and sensitive to initial and current conditions, mereological analog control (from the constraint regime of multiply realizable types to their components and behavior) can produce behavior that is more responsive to local circumstances, moment to moment. As illustrated by analog thermostats and antilock brakes, analog-coded governing constraints of continuous constitutive regimes can implement nuanced and fine-grained behavior, in real time and in a task-appropriate manner.

In summary: (1) coherent interdependencies are described by order parameters that take continuous values. (2) This suggests that there evolved a form of analog decision-making and output control in terms of continuous values.[8] Because order parameters in terms of which output is regulated are contextually generated, analog top-down control is also more timely and responsive to locally changing circumstances. (3) Consequently, analog output control can implement standing causes even as it adjusts to changing current conditions. It can also generalize because it can accommodate learning. Because they are more coarse-grained, continuous values are also more tolerant of error and therefore more robust and resilient than digital ones—as well as more energy efficient.

We turn to the idea of energy management next.

Analog Control and Energy Management

This section proposes that the need for better energy management also favored the evolution of analog governing constraints. Complex ongoing monitoring and regulation of functions like homeostasis, not to mention the deliberate planning and execution of long-term projects in complex social settings (which would become energetically prohibitive with only digital control), are facilitated with analog decision-making and control mechanisms.

British American mathematical theorist Freeman Dyson maintained that for living things to survive, control must be analog: "Controlling day to day or millisecond to millisecond behavior in the real world" would be crippled if (an organism) had to rely on digital mechanisms (F. Dyson 2001).

Freeman's son, George, who inherited the analog mantle from his father, caricatured digital processing algorithms as "DO THIS with what you find HERE and go THERE with the result" (G. Dyson 2011, 277). Digital mechanisms must exactly specify HERE, THERE, and WHEN; consequently, they are precise and fast but unavoidably brittle;[9] they can neither generalize to other domains nor tailor specific responses to altered circumstances. They cannot learn. And because they must precisely control all components and processes through millions of individual and absolutely accurate instructions per second, digital mechanisms also consume immense amounts of energy—to wit, today's server farms.

Most computers are still digital. Today's supercomputers can perform one quadrillion operations per second. It is predicted that data centers, the internet of things, 5G connectivity, and other massive users of energy will consume up to a fifth of the world's electricity by 2025.[10] Speed compensates for energy usage, but about fifteen years or so ago, silicon-based computers began to approach their efficiency limit despite improved techniques that pack more transistors into each chip. By way of contrast, it is estimated that our brains compute around 10 quadrillion operations per second (https://computers.howstuffworks.com/worlds-fastest-computer.htm), and although the brain is by far the body's largest energy consumer, it is in no danger of overheating. Nor does it waste energy. How do our brains manage that feat?

"Analog decision-making processes are far more robust when it comes to real time control" (G. Dyson 2019). Dyson son notes that today's search engines and social networks are centered on the topology of the resulting networks and the pulse frequency of connections. "These analog networks *may be composed of digital processors, but it is in the analog domain that the interesting computation is being performed*. . . . [Social networks] encode and process information in terms of connection frequency and network architecture, what connects to what, the rate at which they connect" (G. Dyson 2011; emphasis added), the order in which they connect, and so on.[11] These parametrized variables reveal the information that matters to the search engines and social networks. Such variables are continuous in the sense of embodying multiply realizable types; actual tokens of types can be more or less typical of an exemplar of which they are realized instances.

Dyson's comments suggest that, in living things, selective pressure for effective input and output evolved analog decision-making and control processes, including analog codes, filters, and interfaces. Analog algorithms can be caricatured as, "DO THIS with the next copy of THAT that comes along," with THIS and THAT encoded analogically, not as discrete and precisely localized signals or realizations but as templates that capture a smeared-out pattern of relationships. As an example, Dyson notes that in the genetic code, which is analog, "*That* refer[s] to a molecular template that identifies a larger, complex molecule by some smaller, identifiable part" (G. Dyson 2012, 177).[12] Templates, as we saw, embody constraints that represent generic patterns or exemplars. In our terminology, they represent type-identified parameters whose token realizations can be specified contextually, in different material substrates, moment to moment, by more local constraints, in a range of instantiations. It is the relational information bound together as a general pattern or exemplar and captured with a new code that matters. Nutrient medical lattices mentioned earlier regulate the rates and pattern of bone growth in this manner. The relations among orientation, magnitude, and growth rate are the variables that matter in bone growth. This templating-by-lattice embodies an analog constraint process calibrated to an attractor's parameter space. Analog-formatted governing constraint regimes with analog output underpin contextually nuanced regulation and control.

More to the point today, Dyson notes that even in social media, "Streams of bits are *being treated as continuous functions*, the way vacuum tubes treat streams of electrons or neurons treat pulse frequencies in the brain" (G. Dyson 2012, 280; emphasis added). As we have seen, frequencies, rates, and topological characteristics are systemwide features that take continuous values. It is in terms of these continuous variables—the unified Ones or interactional types—that "the interesting computations" are performed. By *interesting computations*, read constitutive and governing constraints exercising control on behavioral output in top-down cascades of multiple constraint satisfaction. Solutions that satisfy all constraints need only be located in the feasible, appropriate region. Processing data in terms of those constraint regimes lowers the rate of internal entropy production and delays dissipation. Doing so enacts meaningful and task-appropriate information. Constraint-programming languages and libraries are now widespread.

"Analog is back and here to stay," writes George Dyson.

Analog-coded decision-making and control operate in terms of continuous properties that capture unity of type, a feasible region in state space. Because they span multiply realizable and continuous values, analog-controlled output can be continuously adjusted so that, overall, the order parameters of those types are satisfied and decision-making is carried out in energy efficient and flexible fashion.

This is not to disparage the remarkable advantages of digital processing, speed, and precision. Having information arrive quickly and error free at a decision-making node affords significant adaptive advantage. Accordingly, the initial stages of vision enact digitalized transformations. The retinotopic map in the brain's primary cortex is a paint-by-the-numbers sheet: each point on the map corresponds one-to-one to a point in two-dimensional space. Here, digitalization ensures quick transmission; it also makes quick one-off responses possible. Neuronal spikes are also digital: crossed an activation threshold? Fired? Run! Otherwise, do not. Neuronal thresholds and spike are on/off switches and signals, all or nothing.

That said, there are drawbacks to digital processing and control: converting an analog world to digital form gains crispness and speed but loses depth and richness; it loses the contextual nuance of relations, in other words. Reconverting the digital record back into analog form does not recover that loss, as fans of long-playing vinyl records can attest. Reconversion will not bring back contextual—and qualitative—subtleties present in the original analog world and lost in digital conversion. The elder Dyson joked that this is one reason he would not want the contents of his mind uploaded to a digital computer: "Downloading human consciousness into a digital computer may involve a certain loss of our finer feelings and qualities. That would not be surprising. I certainly have no desire to try the experiment myself" (F. Dyson 2001).

Empirical evidence confirms, however, that information processing and decision-making even in the human brain do not happen in digital format alone. Neural spikes have an analog aspect: the amplitude and duration of those spikes encode nuanced distinctions. Speed is critical, but effective, long-term sustenance and context sensitivity of complex projects are equally critical for decision-making and output control in complex social environments where relations are the variables that matter. It is reasonable to speculate that analog formatting evolved in response to selection pressures. And so, the brain also relies on reverberation, resonance, oscillation, phase, amplitude, frequency, waveforms, and other analog-coded configurations to process information and issue and control behavior. Complex waveforms, neurotransmitters, and hormones in the brain and the gut are evidence that analog control never went away. As an added bonus, analog control is better at managing the brain's computational demands without overheating. No doubt this is because, as we have seen, constrained energy flow that reduces internal entropy production best delays the process of equilibration by organizing into types. Processing in terms of regions is cheaper than processing in terms of points. Degenerate and continuously realized patterns, designs, and configurations persist and remain coherent, sensitive to context, and timely despite internal and external change. Again, without overheating.

To paraphrase George Dyson, the real question is whether organic processes that reformat and recode information—and produce complex behavior—in terms of relations that matter are analog. It appears so. We saw it at work in homeostasis. The brain, too, it seems, encodes higher-order properties not only in its discrete neural spikes but in terms of their amplitude and duration. Frequency histograms of neuronal spike trains (the time intervals between spikes of a single neuron), the timing profile of neural spikes, and other features of neural sequences encoded as analog patterns also transmit higher-order information (Humphries 2021). With respect to medical, legal, and ethical issues, sensitizing the learner to the nuances of the current context through apprenticeships and experiential learning is as important to acquiring new skills as declarative knowledge.

The brain, that is, is both analog and digital. The farther from sensory nerve endings, the more integrative and interpretive its association nodes become, until they reach the prefrontal cortex (PFC), where analog becomes predominant. Early on, the eardrum processes information in analog form: it converts analog soundwaves into analog vibration patterns. Only later are those vibration patterns converted into digital nerve spikes. Even then, higher cortical centers such as the PFC that interpret those digital spikes subsequently reconvert them to analog format.

The PFC modulates executive functions such as planning and decision-making. It is the site where personality and character come together and complex cognitive actions are planned; it is where convergence of already highly processed (read integrated and reinterpreted) information, including emotional, sensory, and cognitive information, is further recoded in terms of social expectations. These capabilities are almost in their entirety encoded in analog form.

Decision-making and behavior control of higher-order information in human beings, in other words, are often conducted by coding and recoding, integrating and reintegrating internal and external signals into increasingly general types of processes held together by analog-coded interlocking constraints. These transmit meaningful path-dependent information; they incorporate history and time. They remember, but they can also be modified by current context and as a result of learning. It is only because decision-making and control processing are analog that, in context-dependent, timely, and appropriate manner, organisms can increasingly regulate and direct attention, inhibit impulse, remember and anticipate, and, especially, act in a more flexible and long-range manner (Wikipedia).[13] Without burning out.

Multiply Realizable Domains Are Analog Spaces

We close with an example on how analog decision-making and top-down control by unity of type—from coherent One to actual tokens—might have evolved.

Survival of the fittest (the idea of one exquisitely fit phenotype uniquely adapted to a fixed and pre-existing niche) is a myth. Interpreting natural selection in this manner implies that evolution tends toward one-to-one—digital—relations between genotype and phenotype, as well as between phenotype and a niche to which the organism is perfectly tuned. This logic allows no room for modification and adaptation in case of significant perturbations (other than accidentally, by random mutations). There would be no principled way to generate resilient and flexible tokens of a general and coherent type. There would be even less room for learning and evolution.

Earlier sections of this chapter proposed that once transitions from many to One occur, effective output is modulated and regulated in an ongoing manner such that it remains within a constitutive attractor's basin—true to type. Contextually constrained interactional types that are analogically formatted and multiply realizable build in the possibility of adaptation and evolution.

The evolutionary hypothesis of "sloppy fitness" can accommodate this perspective.

Agosta and Brooks (2020) argue that niches and habitats are *sloppy fitness spaces*. Sloppy fitness spaces imply the presence of variants. It is important to reiterate that the term *fit* and its cognates were used during Darwin's day to mean a process of mutual adjustment, something akin to a tailor fitting a customer with a new suit of clothes. Fitness space is called sloppy, then, because it can be differently realized under different conditions and yet persist as regulated by a lineage's governing constraint regime, the species' unity of type. Fitness solution space spans a feasible region, not a single point. We saw that process at work in Crithidia's overlapping habitat with plant-eating insects. Nature organizes fitness regions with enough room to maneuver a timely and appropriate fit. In human beings, knowledge is the cognitive, affective, and behavioral organization of concrete experiences into context-dependent and context-generated types and patterns. We would expect this to manifest as a constrained area in neural state space, not a particular neuron or even neural pattern.

Sloppy niches and habitats are therefore co-generated interdependencies that support token variants in a range of contexts. Thinking of niches and habitats in terms of mereological interdependencies explains how *unity of type* can be preserved even while token specimens are continuously tailored and specified to the changing context. These considerations imply that sloppy fitness space must represent a population-level and analog-controlled constraint regime. A type-identified region in state space. Ex hypothesi, the scope and reach of a species' type-defined fitness or possibility space (its potential phenotypic landscape) are always more expansive than the realized state space of particular organisms (Brooks and Wiley 1988, H_{max}—H_{obs}). Cognitively this means that once the toddler grasps the concept of dog, she can recognize a breed she has never seen before as a token of that type. The multiple realizability of contextually constrained interactional types underpins this capacity.

Generating fitness spaces that are flexible, multiply realizable (sloppy), and governed by analog constraints extends adaptive space, the capacity to regulate and control realizations, top down, in real time in response to context. It allows the fitting-together process to be continuous and therefore responsive and tailored to context. By locating decision-making and output control in analog-encoded constraint regimes constructed in terms of unity of type, both a niche's fitness space and its actual tokens will possess sufficient and endogenously produced wiggle room to fit together a richly complex habitat. Ashby would approve.

We can therefore join the Dysons in hypothesizing that over time, coherence-making by constraints evolved layers of integrative dynamics with distinctive forms of selective output control. In virtue of multiply realizable, context-dependent constraints, emergent self-organizing, self-maintaining, self-constraining, and self-determining dynamics there evolved new mechanisms of analog processing and output control. This form of processing and output control supervenes on precise and accurate digital input–output relations (see chapter 13 on supervenience).

With a richer range of options at its disposal, analog processing and control in terms of *tagged qualia* can also underlie increasingly complex behavior. Qualia and phenomenal awareness have survival value, in part, because of their analog character. Police officers often offer analog-coded advice, "Trust your gut. If the situation does not look or feel right, it probably isn't." "Looking and feeling right" is an eminently contextual and analog-coded constraint whose valence has been tagged with a gut feel. Phenomenal awareness of scenes coded in analog form—as multimodal gestalts tagged with a particular feel and processed as affordances or warning signals—allows subjects to scan a whole panoply of sensory, affective, and cognitive information simultaneously and faster. And act on the complex scene appropriately. "I can recognize it when I see it." Nobel economist Daniel Kahneman's distinction between fast and slow thinking might correlate with digital and analog-controlled physiological processes.

Moving decision-making and control to the level of constraint regimes (that embody type-level properties and are tagged with phenomenal awareness and qualia) saves energy and avoids overfitting in the process. By avoiding overfitting, the logic of type-defined possibility spaces formed by contextual constraints and governed with analog control confers a selective advantage on its token specimens.

Finally, what the elder Dyson refers to as our "finer feelings and qualities" may well be those higher-level affective and epistemic properties with analog control powers that emerge from the translation of discrete signals to an analogically coded and tagged set of interdependencies. Such multimodal integration and entrainment into a complex and analog attractor with emergent properties, including novel forms of phenomenal awareness and qualia, might be exactly what allows wine connoisseurs to perceive a far richer, relational, and context-dependent range of interdependent characteristics (tagged phenomenally as a wine's terroir, for example) with reference to which they recognize and evaluate vintages.

We can conclude that, over evolutionary time, constraints evolved novel codes that transduce Many-to-One transformations into newly organized constraint regimes with the continuous properties of emergent order parameters. Tagged as qualitatively new properties, these governing constraints embody mechanisms of analog processing, decision-making, and top-down control.

* * *

English anthropologist Gregory Bateson (1960, 1972) might have been the earliest thinker to note that "alternating digital and analog levels" is typical of hierarchical organization. Because hierarchy theory is a theory of differences, not sameness (Salthe 2012), it is ideally suited to explain qualitative differentiation that arises from constraints. The next chapter examines how Many-to-One transitions create hierarchies with asymmetric and qualitatively distinct levels of organization that interact in terms of constraints, not forces.

12

Of Holons, Holarchy, Heterarchy, and Hierarchy

Koestler's Holons and Holarchies

Intent on studying relations between individuals and societies from a realist perspective, Hungarian British journalist and author Arthur Koestler (1967) coined the term *holon* to describe compositional or nested relations. He depicted holons as Janus-like. Nation-states, for example, face inward toward those persons, groups, and institutions that compose them. They simultaneously face outward toward the world in which they are situated and from whose perspective a particular nation-state is one of many. In our terminology, holons are constrained and constraining interdependencies between parts and the wholes they comprise as well as between those wholes and the context in which they are embedded. Complex systems such as dissipative structures are holons (Pattee 1973; Salthe 1985; Eldredge 2015).

Coherent dynamics generated by constraints are also *holarchies* (Koestler 1967) and *heterarchies* (McCulloch 1945). Like *holon*, these two terms were coined to differentiate the concepts involved from the common connotations of hierarchy. Holarchies are relations of interdependence whose components are themselves holons. Holarchies integrate upward to more encompassing holons; they exercise control downward, on their component holons and behavior. Heterarchies are relations of interdependence in which no one unit or level controls the rest; unlike the classical understanding of hierarchy, heterarchies explicitly allow influences between entities at the same level of organization. This book uses the more common term *hierarchy*, but without the connotation of an absolute top and bottom; constraints, as explained here, can have effects at the same level of organization. As used here, therefore, *hierarchy* also refers to holons, holarchies, and heterarchies.

Consequently, rehabilitating the concept of hierarchy requires some clarification about what these coherent dynamics, so understood, are not. In business and the military, the term *hierarchy* often recalls inflexible, stratified legacy organizations where recommendations from lower echelons are systematically ignored by managers and officials.[1] On this standard interpretation of hierarchy, inflexible and one-way, top-down force and power relations are the defining characteristic of hierarchies—so much so that *rigid hierarchy* comes across as a redundant description. Since mereological constraints play a significant role in this work, hierarchical control as understood here underscores why Chinese boxes and matryoshka dolls are not proper hierarchies: they are mere containers; they exert neither enabling nor constitutive constraints on either their contents or their embedding contexts.

Governing constraints and hierarchical top-down control in complex systems are best conceptualized, therefore, on the analogy of ecosystems, synchronized clocks, and the invisible hands of economic systems. Whether referring to organisms or human organizations, hierarchical control as understood here is like the influence of virtual governors that emerge bottom-up from interactions among components. As described in previous chapters, virtual governors are distributed interlocking constraints that exercise top-down control on their components and behavior. They describe an ecosystem's regulation of rates and ratios of flora and fauna. Or the way a firm's culture and tacit rules frame employees' attitudes, behavior, and dispositions, which change qualitatively as a result of the culture's overarching constraint regime.

This way of thinking of hierarchical control rejects the notion of control hierarchies as the arbitrary top-down dictates of rulers and other public officials. For that reason, examining the use of raw, top-down political power to coerce and control individuals or groups is beyond the purview of this chapter. Although this is a real and serious threat to humankind that requires addressing and correcting, it is best understood as the exercise of force, not of constraints as understood here.

In short, the previous chapters proposed that natural hierarchy formation and control are the outcomes of constraints. They are coordination dynamics generated by enabling constraints against a backdrop of context-independent constraints; interlocking constitutive constraints serve as a coordination context that modulates, controls, and regulates components and behavior top down—as analog governing constraints. Contextually dependent interdependencies are therefore holarchically organized regimes of constraint (Allen, O'Neill, and Hoekstra 1984).

Ontically, they are spatiotemporal interdependencies that constrain and are constrained by their contextually intertwined components, as well as by the constraints of the context in which they are embedded. Although intralevel holarchical influences also generate these interdependencies, this chapter focuses on mereological relations *between* levels of hierarchical organization in contextually coherent dynamics.

* * *

Recognizing that systems of constraint are organized hierarchically is not news. However, because of the received understanding of mereological relations philosophers have rarely focused on relations between processes at different hierarchical levels. Hierarchy theorists like Pattee, Salthe, and Allen and Starr who regularly emphasize the role of constraints in relations between hierarchical levels stand out for that very reason.

Hierarchy theory provides two entrées for the approach presented in this work: (1) relations between levels of hierarchical organization are not energetic exchanges; they are relations of constraint. This is particularly significant with respect to top-down control from whole to parts. (2) Levels of hierarchical organization ordered by contextual constraints cannot be extensionally defined in terms of their reference. Properties and powers characteristic of each level are generated by specific enabling constraints. History plays a leading role in these constraints, as do spatiotemporal constraints generally. The properties and powers of contextually generated dynamics must therefore be defined intensionally, that is, in context. (3) Finally, philosophers have long puzzled over the power of the content of mental events and intentions to bring about actions that satisfy that content (Juarrero 1999). As characterized in chapter 1, those debates were framed by a flawed understanding of mereology and causation. Coherence-making by constraints, reframed in light of hierarchy theory, allows the two to find common ground, as this chapter will describe.

* * *

Less than two years after Koestler published his treatise on holons, Nobel laureate Herbert Simon (1969) described the stabilizing benefits of hierarchical organization. Arguably, Simon's lecture formally launched hierarchy theory as an academic field of study. Although his interests were centered on computing systems, he primarily addressed the contribution of modularity to hierarchical organizations.

Simon presented the watchmaker analogy in his now-famous MIT Compton Lectures. Consider two watchmakers who manufacture pocket watches composed of one thousand pieces each. The first watchmaker crafts a new pocket watch each day from a supply of hairsprings, case screws, wheels, jewels, pinions, and so on. The second assembles modules of one hundred components each. Assume each watchmaker is regularly interrupted after 150 elementary parts have been assembled. Interruptions cause any components "that do not yet form a stable system" to fall apart. Under these conditions, the first watchmaker will be hard-pressed to complete a watch at all; not so the second watchmaker, who can rely on stable modules to stay on track. On average, this second watchmaker will manage to complete one watch after every eleven interruptions. Using modules in assembling a complex object prevents backsliding and having to restart the assembly from scratch (Simon 1973).

In contrast to individual springs and screws, modules are complex arrangements of interdependencies whose integrated processes and parts realize stable units. As mentioned earlier, modules are condensates of constrained interactions (Burgauer 2022a). Modules are therefore holons, assembled of individual and separate elementary parts with a view toward the extended interactions in which they are embedded.

Biological degeneracy is often realized through modularity; modules embody different paths along which the same function is actualized, a capacity made possible by the multiply realizable governing constraints of a system's overarching dynamics. Neurological regions subtending cognitive function are eminently degenerate. But modules need not be physical: subroutines in computer programming are modularized constraints. Plug-and-play modules are not just extra copies of a part. Material or not, modules are mini holons organized with a view to the realizing role they play in more encompassing holons. They can be materially different and yet can be plugged in and played without additionally reconfiguring the entire system.

One module can be replaced by another as long as it realizes the same effective input to a coherent structure. For example, watch escapement modules include a bearing on which the escapement wheel rotates; their qualitative character as escapement modules (an emergent property) is conditional on this friction-reducing property. In the late 1960s, low-cost quartz escapement modules replaced the classical and more expensive jewel-based ones; this was possible only because quartz is in the same equivalence class as rubies with respect to friction reduction, an order parameter. The overall watch's interdependencies, that is, filter or interpret quartz modules as effective input just as well as jewel-based ones. Slightly different tokens can likewise replace functionally equivalent ones. Human organs can be replaced with artificial ones because they are functionally defined and regulated modules.

Modules counterbalance the conservative role of entrenching constraints. As the discussion on entrenchment showed, modules stabilize assembly processes, but as the escapement module illustrates, they also prevent systems from crystalizing into one rigid and brittle configuration. By securing one stage or level of hierarchical organization while simultaneously allowing other pathways to carry out the same function, modular flexibility supports easy repair and modification; modules prevent ongoing coherent interdependencies from distintegrating and decohering. In other words, modules encapsulate effective enabling constraints that facilitate the generation of resilient hierarchies.

Constraint-generated coordination dynamics therefore lend themselves particularly well to modularly organized hierarchy formation. Although the term *hierarchy* often connotes spatially nested configurations, temporal constraints also play a central role in hierarchy formation. As has been described, integration of separate entities into holons can be the product of temporally coded enabling constraints. The evolutionary sequence of vertebrates, mammals, and primates describes distinct stages of a hierarchy regulated by temporal constraints. Each stage is a stable module from which future specification proceeds (Salthe 2012); individual stages simultaneously constrain future steps and are constrained by earlier ones.

* * *

In contrast to the classical concept of hierarchy that assumes an organizational top and bottom with control operating exclusively top down, holons, heterarchies, and holarchies are always tripartite or trilevel. As a contextually constrained coordination dynamic, each is embedded in (not just plopped into) a more encompassing holon whose governing constraints establish the embedding framework within which the focal holon or stage is situated and which regulates and modulates its components. This embedding context is often taken for granted, which is why hierarchy theory underscores its significance. Most early hierarchy theorists were biologists, where context reigns; some (Weiss 1971) expressed dismay at science's continuing failure to take context into account.

Although scientific observations and methodology are naturally centered on a particular level of organization, Salthe (1985, 2001, 2012) systematically keeps the three hierarchical levels of developmental processes front and center in his writings. The holon of interest is the *focal level*. In addition to components, which comprise its lower levels, the focal level is also always embedded in a more encompassing holon, the structured context whose emergent properties and powers arise from the constrained relations among the entities at the focal level. Significantly, the context's properties and powers are different from those of the focal level and its own embedding context. They are qualitatively different because they are relational; they reveal novel properties that emerge from the constrained interactions and relations among the components. They are real despite being relational. Western philosophy and modern science have taken for granted the ontological primacy of simple particles with primary properties. This assumption was first called into question by the entangled context-dependent interactions at the quantum level. Advances in our understanding of ecology have brought context to the forefront.

Despite their context-dependent texture, Koestler noted that, as units, holons exhibit self-assertive tendencies that preserve and advance their individuality and quasi-autonomous character (Koestler 1967, 343; Allen and Starr 1982). Chapter 7 showed that feedback loops as well as closure of processes and constraints generate such coherent dynamics; the assertiveness Koestler describes is due to the enhanced metastability of those interdependent dynamics. Over time, they become autonomous and self-determining. But even in medicine, homeostasis's governing constraints can compensate and keep the organism metastable as an identifiable coherent unit despite spleen and gallbladder removal, for example. Having component modules that are themselves holons strengthens and prolongs robustness and resilience in the abiotic and biotic domains alike.

Koestler's anthropomorphic term *self-assertiveness* is therefore best replaced with *robust coherence*. It is because of the enhanced metastability conferred by interlocking interdependencies among constraints at many scales that modules help watchmakers finish the job. Modules in nonliving and living things alike provide multiple paths to the realization of the emergent interdependencies that constitute a type of structure or function; this metastable flexibility, in turn, supports the integrity, resilience, and persistence of holons over time.

Koestler's neologism was short-lived, but as this work has repeatedly pointed out, the minimally trilevel or triscale character of naturally formed hierarchies is central to coherence-making by constraints. Allen and Starr adopt Koestler's terminology; however, they consider holons to be solely epistemological constructs, "integrations of a set . . . and sets are a matter of ad hoc definition" (Allen and Starr 1982, 10). Here, in contrast, we presume that constraints that generate and maintain these interdependencies—and the interdependencies themselves—are real and context dependent.

Levels of Organization

Watch modules and anatomical structures (cells, tissues, organs) show how some hierarchies can be treated as "nearly decomposable" into distinct levels of organization (Simon 1969). However, this matryoshka doll image of stratified units is misleading. As mentioned earlier, software subroutines, hypercycles, closure of constraint, and iterative recursion can confound the clean material separability of layers of constraint. Consequently, hierarchy theorists often analyze decomposability dynamically, as differences and changes in phases, rates, scales, amplitude, timing, and other such order parameters. Levels of hierarchical constraint dynamics such as cells, tissues, organs, or individuals, groups, and organizations for example, are separated by distinctive rates of processes that "continue, cycle, or go to completion at very different average rates when viewed from a fixed scale" (Salthe 1985, 72). Such differences and changes present as identifiable hierarchical levels (Pattee 1973; Allen and Starr 1982; Eldredge 2015). In a multidimensional phase space, all can coexist.

Boundaries between levels of organization and constraint are therefore typically marked by sharp discontinuities in order parameters and average process and relaxation rates. Because they constitute more encompassing and streamlined Ones, overarching hierarchical levels of organization are usually slower than their components or realizations. The smaller the "average rate constants" that define a set of interdependencies (Salthe 1985, 72), the higher its organizational level. Stringent coupling between tokens of the same type is often implemented through shorter feedback loops. Prey interact more frequently with conspecifics, that is, with others that share the same set of interdependencies, than with predators. Tight feedback loops among members of a business development team help behaviors and processes coalesce as a unit. Fast iterations are also more responsive to current context because feedback from the environment quickly updates the conditional probabilities of the constitutive regime's interdependent constraints with respect to the environment. By way of contrast, sharp discontinuities to longer feedback loops and governing constraints with a longer reach suggest a many-to-One transition to a more encompassing hierarchical level or to an earlier form of organization, development, or evolution of which the more recent tokens are specifications.

Rephrased, embedding interdependencies are typically slower and longer continuous constraints than embedded ones, which the former smooth over. Ecosystem coordination dynamics, for example, are coarse-grained, longer and slower than those of its individual species.

* * *

Relations between the faces of Janus, from focal level inward and downward to its components, and outward and upward to its embedding context, are negotiated at permeable boundaries that mark these discontinuities. To repeat, these boundaries or transition regions should not be thought of as exclusively gaps or walls; like eardrums, they are interfaces, active dynamic gates that recode, standardize, harmonize, and adjust rates, phases, amplitudes, and other properties between one level and the next, between each holon and the one lower or higher. Such processes of constraint satisfaction are regulated by those interdependencies that mark the holon's coherence and self-assertiveness. Acting as filters, codes, and gates between levels of organization, interfaces effectively sift, integrate, translate, transduce, and refract elements and dynamics at one phase or rate, but always in reference to constraints, elements, and processes at other levels. The constraints that define the interface adjust rates, timings, bufferings, and so forth such that the multiple constraints on each side are satisfied and the integrity of the whole remains both coherent and open to the environment.

As just noted, differences in strength, manner, speed, and degree of coupling commonly indicate the presence of a boundary or an interface between hierarchical levels. Degree of coupling generally shows sharp discontinuities with distance (temporal or spatial) and strength. In physics and chemistry, elements belonging to the same set of interdependencies share similar bond strengths. Bonds that hold nucleotides together are stronger than those that bind electrons to the nucleus. Intermolecular forces are much weaker than intramolecular covalent bonds. "It is precisely sharp gradation in bond strengths at successive levels that causes the system to appear hierarchic and to behave so" (Simon 1973, 9). Individuals tend to be less close with friends who now live far away than to those with whom they interact daily.

Constraints that result in tight coupling often generate local organized interdependencies we call clusters or hubs. Network hubs and cluster analysis of linkage relations like those of terrorist networks can reveal unmistakable evidence of the correlation in degree of information exchange, social coupling, and authority relations among its components. These parameters represent an encompassing hierarchical level whose interdependencies take longer to achieve closure. It becomes a new context whose rates and cycles streamline and smooth over the faster, lower-level processes entrained into that new holon. Network analyses can distinguish tribal cultures from urban ones by identifying the enabling and governing constraints that constitute a network's profile.

To be clear: this work does not question that nuclear and molecular forces and bonds, for example, involve energetic exchanges. The goal of this chapter, rather, is to call attention to the significance of constraints such as rate and timing differences between inter- and intramolecular forces and bond strengths—that is, to call attention to constraints, boundaries, and relations of dependence and influence between levels of hierarchical organization. Interlevel relations are negotiated by filters, gates, lenses, interface settings, governing constraints, and so on. All of these bring about changes without adding energy or being themselves permanently changed in the relation.

The faster the dynamic, the finer the filter; the slower the dynamics, the coarser the filter. The faster the dynamic, the narrower the criteria that control the lens or gate aperture; the slower the dynamic, the wider the criteria. The terms *filters* and *lenses*, of course, are metaphorical for constraints. Tissues, for example, are continuous and smoothed-over, coherent and type-level interdependencies; they embody and enact multiply realizable regimes of constraint that are smeared out in space and last longer than those of their discrete and separate component cells, which view the context in which they are embedded as a constant. From the perspective of the organ, changes in the tissue are irrelevant so long as their characteristics, translated into effective input, can be smoothed over as supporting the organ's function. If changes in the tissue are too small and fast, on the other hand, they can escape the organ's notice (Allen and Starr 1982). A tumor results.

Because of their context dependence, parameters of control hierarchies are continuously adjusted through processes of multiple constraint satisfaction. Iterative and recursive processes tighten or loosen coupling, speed up or slow down rates, entrench or relax rules that control timing and phase, and so on. New information from current interactions with the environment becomes folded in to those adjustments. Conditional probabilities are updated and activation and inhibition settings are modified by cycles of multiple constraint satisfaction, without directing intervening energetically in the process. Nevertheless, the system's constitutive and governing constraints are preserved as a result. Its dynamic—whether biological or social—remains true to type and endures.

Recent research on inflammation is particularly interesting in this regard (Medzhitov 2021). Homeostatic equilibrium depends on a normal range of background constants such as ambient temperature and adequate nutrition. Activation and suppression of inflammation is always in the service of homeostasis. For that reason, credible and immediate threats of starvation or frostbite can trigger anticipatory inflammation even in the absence of structural injury or pathogens. Threatened loss of homeostatic regulatory control, that is, can directly produce responses that restore regulatory control. Specifically, these responses include behavioral changes such as acting sick—lethargy, sleeping more, eating less, and so forth. Might obsessive-compulsive and autoimmune disorders, post-traumatic stress disorder, paranoia, and depression arise from similarly malfunctioning governing and constitutive constraint regimes that have lost their capacity for regulatory control? These might be cases where the modulating dimmer switch did not dial down. The central point here is that such pathological behavior can occur, not only in response to actual and meaningful physical signals but also as a result of malfunctioning or weak governing constraints—that is, from overly entrenched governing constraints that failed to update its interdependencies to current interactions with the environment. The thermostat itself malfunctioned.

Mereology Revisited

In short, constraints operating at one hierarchical level do not directly transfer energy to entities and processes at other levels of organization. This is especially significant with respect to whole-to-parts or top-down control. It is important because causality by constraint allows hierarchical control[2] constraints to implement top-down causation without falling victim to overdetermination or violating causal closure.

Regulatory genes, for example, control the expression of genes that code for proteins that repress operator genes. The interface between regulation and operation modulates outcomes top down by constraining the timing, levels, conditions of activation, and inhibition of structural genes. Regulatory genes implement modulation not by actively participating, for example, in actual protein construction the way structural genes do, but by controlling the timing, location, and amount of gene expression. They regulate as second-order top-down constraints.These processes of multiple constraint satisfaction are carried out as changes to the possibility space and to the probability distribution of events within it, with relations between hierarchical levels of organization consisting of coevolving process of constraint satisfaction. Bottom-up, independent matter, energy and information flows are integrated by enabling constraints. The resulting coordination dynamic becomes a new and embedding hierarchical level of organization, a new context. Those embedding interdependencies in turn then control their embedded components through ongoing second-order top-down constraints that keep the coordination dynamic whole.

* * *

In a witty illustration of top-down hierarchical control by constraints, Allen, O'Neill, and Hoekstra (1984) describe a situation all too familiar to academics. No matter how heated a discussion among faculty members might become, how loud the voices, or whether they are in the same department or not, in the end, they return to their offices unperturbed—unless one of them is a department chair or dean with authority over others. Rate-dependent energy transfers are always possible—that is, forceful disruptions like yelling and even fisticuffs might break out—but no control constraints are at work among individuals at the same hierarchical level.

Stronger and more frequent interactions among members of the same academic department than between professors and administrators confirm hierarchy theory's axiom that processes at a given level of organization are best identified by their "significant mutual interactions" (Salthe 1985, 23). As noted earlier, strong coupling and autocorrelation among certain processes and entities suggest they belong to the same hierarchical level. In contrast, looser coupling and fewer interactions suggest relations with processes at other levels; but it can also mean that interdependencies at a given level are not well organized (compare tropical storms with category 4 hurricanes). Different hierarchical levels of organization such as the level of type-defined interdependencies and that of the tokens that specify them relate to each other as constraints.

As Salthe puts it, dynamic processes at one level cannot interact energetically with processes at another level (Salthe 1985, 73), if nothing else "because [faster] lower-level dynamics would have been completed before feedback from the [slower] upper level arrived." We saw it at work in constraint closure; hyperloops take longer to complete than the individual catalytic reactions of which they are composed. So too with administrators and faculty members; they are differentially defined and constrained by the rules, timing, scope, and reach of the various rights and responsibilities of their respective roles. The privileges and duties of each role go to completion at different rates: department chairs prepare annual budgets, deans supervise seven-year tenure evaluations and university presidents and chancellors oversee multiyear capital campaigns and ten-year accreditation procedures.

That these constraints are organized as relations of dependence is evidenced by the top-down control administrators exercise on junior faculty members: they can speed up or delay tenure-awarding procedures, for instance. In turn, top-down constraints on a tenured professor's capacity to act might take the form of providing or restricting access to travel funds and teaching assistants. None of these modes of influence is a direct kinetic interventions in an affected trajectory (Salthe, personal communication 2021). Each is a deployment or removal of an obstruction or facilitating pathway, a change in the rules of the game, or of timing constraints. In short, introducing or modifying enabling or governing constraints, raising or lowering barriers to opportunities, and modifying pathways that change the playing field whereby duties and obligations, rights and benefits are divvied up are all ways in which interlevel constraints are exercised. Changing the capacity to act, that is, happens by introducing or removing, or tightening and relaxing constraints; transformed constraint regimes reconfigure the possibility space and the conditional probability of the events that carve its topology.

Pointedly, then, mereological relations among processes and entities at different hierarchical levels are relations of constraint, not efficient cause. Changes in constraints at one level bring about changes in properties and powers at a different hierarchical level, bottom up as well as top down. As an example of tacit bottom-up constraints at work, a dean or department chair's usual top-down control over faculty members is weakened if the latter can threaten to take their multimillion-dollar grants to a competing institution. These relations do not describe energetic interactions, but when those indirect enabling constraints appear, they blur hierarchical levels and top-down control is muted or lost—all without any extra energy having been expended.

Relations from whole to parts are therefore "rate independent" (Salthe 1985, 72) with respect to processes on which the whole acts as a constraint on its components. By *rate independent*, it is meant that, top down, governing constraints "inform and influence" components and realizations without participating in them dynamically (Salthe 1985, 82). The timing of the playground swing kick does not add energy to the kick. More generally, constraints exert mereological control by making the playing field more or less rugged, the gradient of its attractor more or less steep, by expanding or contracting the boundaries of possibility space, positioning a hillock or a chute nearby or moving it far away. A new protocol, for example, can change the topology of possibility space dramatically. Enabling and governing constraints can do this across a variety of dimensions and time scales. These include historical or temporal factors as well as spatial and cultural ones. We saw the roles played by buffers, isolation, and entrenchment. New amendments to a nation's constitution were likewise described as changes of constraint that trickle down to considerable effect.

In general, then, governing constraints of hierarchies control top down by adjusting the constraints by virtue of which the lower level's dynamics are regulated: by adjusting, for example, their sequential order, timing, frequency, activation, and inhibition and conditions under which they can or must happen, or the probability distribution of their realizations. These constraints are recalibrated at the interfaces between the hierarchical levels, by changing their filters and operational rules, regulations, and criteria and therefore, indirectly, the process of multiple constraint satisfaction. When encountering mismatches between outside and inside, or between one hierarchical level and an adjacent one, active interfaces such as cell membranes and eardrums—or divorce mediators—fine tune constraints that control rates, frequencies, wavelengths, and format (digital, analog) of the input, the internal milieu, or both to make them compatible and facilitate interaction. Filtering and selection themselves, as we saw, are processes of multiple constraint satisfaction with a view toward integrability into an effective constraint regime, not energetic exchange. The results: the pancreas releases more insulin (a catalyzing enzyme) in response to the need to manage a fudge brownie and maintain homeostasis. In social organizations, policy changes by the board or an effective leader, or recommendations by an agile coach, can bring about wholesale transformations by modifying the rules of the game.

There are limits to the influence of interlevel constraints. The asymmetry between embedding holons and embedded ones can be destroyed by either excessively accelerating the dynamics of an embedding environment or slowing down the rates of the focal-level system beyond its tolerance range (Allen, O'Neill, and Hoekstra 1984). Think of tolerance range as the boundaries and coordinates of feasible space. As another example from academia: a number of community colleges in the United States now offer four-year degrees; even more offer college credits to students still enrolled in high school; liberal arts colleges have morphed into universities offering graduate degrees. Altering the defining cycling rates of each institution blurs their identities as coherent entities with mutual interlocking interdependencies. It also blurs the asymmetry between them. What was formerly a proper component (feeder?) holon becomes a near sister holon to its former context. When holons are absorbed into their embedding environment, they lose identity through dissolution. That said, some societies are quite skillful at coopting rebels and dissenters and absorbing them into their constraint regime (aka the establishment).

Lack of asymmetry in cycling rates between focal level and the context in which it is embedded, or between the focal level and its components, reflects the absence of control constraint by one on the other, regardless of signal amplitude. It also reflects an absence of qualitative differences between the two levels of organization. For example, when large multinational corporations can shrug off billion-dollar fines imposed by federal courts as the cost of doing business, their formally subordinate relation to the government becomes more like a relation between equals.

In contrast, facilitating energy flow to the focal holon by speeding up its endogenous cycling rates has different consequences. Constraints that enable such speeding up (like shorter loops of positive feedback, a steeper dynamic attractor, or new channels or gradients that deliver more resources to that holon) result in faster autocorrelation rates. Since embedding holons typically smooth over differences among their components and treat them as continuous relations, speeding up rates of events at the component level can make them "invisible" to their embedding context, which no longer detects them as "components" (Allen and Starr 1982).

Examples from oncology are illustrative. Normal cells that start to replicate too fast soon die without nourishment. However, malfunctioning governing constraints on normal blood vessel formation and maintenance might result in the runaway cell becoming vascularized. When this happens, the cell's self-assertive traits can slip the control of its embedding organ. Cancer ensues. Stopping vascularization to such tumors has been shown to resolve the problem as definitively as radiation intervention (an energetic intervention that kills those cells outright). The manner of resolution of the two is different, however. Radiation therapy destroys cells with intense energy, a mechanical force intervention. Cutting off blood supply to the cells, on the other hand, does so by removing an enabling constraint. Differences in the modus operandi of enabling and governing constraints—and between these and mechanical forces—underlie asymmetries and dependence relations between levels of organization, as well as between the organization and its environment.

Allen and Starr (1982) remark in passing that speeding up certain processes in physical chemistry might have led to the origin of life. This new take on the origin of life suggests that overly assertive holons (perhaps due to the influence of autocatalysis, positive feedback, and constraint closure) might have slipped the control of geophysical constraints and acquired sufficient self-determination and autonomy to loosen the physicochemical context's control over them.[3] Such exponential but stabilized growth of coherent interdependencies presupposes the operation of top-down constraints.

As suggested by the prairie grasslands and chaparral examples, Allen and Starr (1982) also claim there are no regular perturbations in nature. Habituation and cooptation modify and adjust ecosystem constraint regimes in response to regular perturbations. In contrast, rare tsunamis and meteorite impacts are true perturbations, one-off events to which coherent dynamics cannot be tuned and to which they cannot become anti-fragile. These perturbations are external blows that impact existing ecosystems as efficient causes. Likewise, extreme cases of invasive species can simply wipe out existing ecosystems; entirely different constraint regimes and token realizations replace them. The damage such perturbations wreak is real, but it is best understood as the effect of forceful impacts, not the workings of constraints.

Rate and phase differences between processes at different hierarchal levels of contextually constrained interdependencies thus reflect qualitative differences in properties and powers at each organizational level. Hierarchical levels of organization capture features that demarcate the kinds and types of entities realized by each set of constrained interdependences. Differences in constraints induce different levels of organization; different levels of organization represent different emergent qualities. Each hierarchical level is a set of interlocking constraints marked by uniquely individuated interdependencies. These become manifest as qualitatively novel emergent properties, powers, and behavior patterns. Constitutive constraints at each level of organization hold together qualitative distinctions that make interlevel relations nonreciprocal and asymmetric. As an example, contextual constraints imposed by domestication facilitated the emergence of qualitative differentia between dogs and coyotes.

Some of these constitutive constraints are more obligate than others; they are entrenched. Others are more facultative and short-lived. For example, constraints that differentiate domestic pigs from wild boars are quite fragile; they hold only within a narrow range of conditions. Within one generation, wild boar traits resurfaced in domestic pigs turned loose in the Florida Everglades. Nevertheless, no type can be understood absent the context and the constitutive constraints that hold the type's coherence together. But given a specified context and the enabling constraints that generated the type, those interlocking interdependencies constitute real (if local and temporary) dynamics with qualitatively novel and emergent properties their components lack.

Extensional and Intensional Definitions

Different contextual constraints with different reach and scope generate qualitatively different interactional types. Tokens of those types embody and enact those qualitative distinctions. That is why meaningful and qualitative distinctions are disambiguated by context and in terms of the constitutive constraints of their embedding context. When context-dependent constraints generate indexicality (here, there, up, down, inside, out, before and after, in these circumstances but not those) qualitatively new properties emerge. These indicate that the coordinates, boundaries, and topography of the possibility space have transformed at multiple scales. Contextually constrained types and their tokens can only be explained in reference to newly emergent relational properties indexed to a particular context at a particular moment.

Hierarchical structuralism is therefore a theory of differences, not similarities, much less averages. But modern science traffics in averages, bell-shaped curves, and eigenvalues. How does science handle qualitative distinctions or individuation, both of which depend on context?

According to modern science and philosophy, terms referring to events and processes must be capable of *extensional definition*. Their definition demarcates their reference. Extensionally defined terms can be substituted in statements without loss of truth value. For modern science and philosophy, as chapter 1 described, extensional definitions were grounded in fundamental particles and anatomical structures; these were assumed to be determined by eternal, unchanging, and universal primary properties. Contextual differences were attributed to secondary properties and were considered ontically derivative. Western science and philosophy to this day still aspire to "the View from Nowhere" (Nagel 1986) and Spinoza's sub specie aeternitatis perspective.

This appealing promise of transcendence, I submit, has a negative downside. It dismisses nature's ontic capability for generalization, individuation, and creativity. In contrast, this book has argued that context is real and contextual constraints produce real effects, among which are creative emergence of new types of entities with uniquely individuated trajectories. When contextual constraints take priority as a system's enabling and governing constraints, providing extensive definitions in terms of internal and universal properties becomes a nonstarter. Emergence and creativity, however, bring with them the unavoidable uncertainty that comes from context dependence's capacity to generate types of entities that are, by definition, indexed to the context and multiply realizable. In consequence, the asymmetry between hierarchical levels of organization (Salthe 1991) formed by contextual constraints bars events, properties, and powers embedded within each level from extensional definition.

Specifically, contextually constrained interactional types (Ellis's equivalence classes; Ellis 2016, 2021) cannot be defined by way of their reference or denotation. As we saw in the discussion on effective input and output, interfaces of complex systems calibrate incoming physical signals in terms of their relevance to the system's constitutive constraints and their order parameters (cohesion, fittingness, vorticity, valence, and so on). In context. Cohesion, fittingness, and the rest of these relational properties are eminently indexical and context-dependent; they subtend the context-dependent powers of those interdependencies.

In conclusion: hierarchy formation by constraints implies that organizational levels and their realizations must be *intensionally defined* in terms of novel qualities that arise at each level of dynamic organization in response to integration by contextual constraints (Ellis 2016, 2021; Salthe 1991).

Many-to-One transitions in response to constraints therefore mark phase changes to hierarchically and indexically organized possibility landscapes where each point in space and time now has a neighborhood and reflects a moment. Points in hierarchical space must be defined intensionally, in terms of the contextual constraints and interdependencies in which each point is situated. It must be defined, that is, in terms of the holon, attractor, or coherent One—in which it is embedded. When context matters, space is no longer a featureless container; it has become a *place* structured and limned by constraints. Points in time become positions and moments, steps and stages. Consequently, context-dependent constraints that were assumed to go without saying must be foregrounded to fully explain the role they play in those rumpled possibility landscapes and dynamics. And therefore in the individuated landscape of each token in it. By focusing on asymmetries and differences in the interdependencies enabled by context-dependent constraints, contextual constraints inevitably implicate *intensional definitions*. Thinking of universally applicable counterfactuals as the defining characteristic of science becomes problematic. Chapter 14 returns to this subject.

Control Hierarchy

Part III brings the backstory to the present day, where mereological and supervenience relations take front stage. The next chapter discusses mental causation by constraints that embody nontransitive intensional content, as suggested by the supervenience relation.

III

13

The Backstory, Today

When we last left the backstory, Descartes had proposed the theory known as dualism, the sundering of reality into two domains, mind and body. As we saw, dualism could not account for top-down causes without violating physical closure or espousing overdetermination. Neither could it explain coherent wholes or intentional, purposive action.

The early twentieth century witnessed the apogee of Skinnerian behaviorism, a reaction to dualism which I will not discuss since it is so well known. By midcentury, however, objections to behaviorism were strong and numerous. Empirical research cast doubts on the idea of mind as the impenetrable and/or unnecessary content of a Skinnerian black box. Attempts to blackbox cognitive processes and reduce sensations like phenomenal feels to operant conditioning floundered (Rosenblueth, Wiener, and Bigelow 1943; Putnam 1967, 1975; Wright 1976; Goldman 1970; Roque 1987). By the mid-1970s, a new academic discipline called *cognitive neuroscience* turned to computer technology and brain imaging, especially attempts at brain mapping, for inspiration on the relations between mind and body.

Formulated by British philosopher J. J. C. Smart (1959) and Austrian American philosopher Herbert Feigl (1958), the *psychoneural identity theory* (also called identity theory) set the stage for this new approach by proposing the reductionist thesis that mental events and processes are identical to—nothing but—brain events. U. T. Place's (1970) seminal article was titled "Is Consciousness a Brain Process?" In language typical of its day, the theory proposed that, for example, "mental event or process Ψ is identical to C-fibers firing."[1] An inadequate understanding of identity, informed by the standard understanding of causality as energetic transfer, was to plague the identity theory as well as the two that followed, functionalism and the supervenience thesis. Let us see how.

The earliest version of the identity theory proposed that mental events generally—my thoughts now about my grandmother, for instance—are identical with a certain neuronal firing pattern. The identity theory allegedly applied to phenomenal feels as well. Pain—my headache now—is also identical to (nothing but) a certain neuronal firing pattern.

Type identity theory held that types of cognitive and sensory processes (such as thoughts, beliefs, pains) are identical to certain types of brain events. Headaches are identical to a type of brain event—say, a C-fibers-type brain process. Type identity would be falsified if two tokens of the same mental event (two instances of headaches) correlate with distinct types of neural or physical events—for example, if my headache now correlates to certain C-fibers firing but your headache yesterday correlated with a different type of brain event, such as a neurotransmitter-release-type event. Since C-fibers firing and neurotransmitter releases are distinct types of brain processes, pain as a type of event (headaches, in this case) could not be identical to C-fibers firing type events.

Evidence to that effect quickly appeared. Degeneracy popped up everywhere (Mason et al. 2014). Dramatic cases of patients with congenital hemibrain but who showed no obvious evidence of cognitive impairment raised serious doubts about any simple one-to-one identity between types of mental events and types of physical events such as brain processes. Neurological studies establishing the indubitable plasticity of the brain put paid to any claims to identity.

To circumvent this objection, the *token identity theory* replaced type identity. This weaker version of identity maintained that actual instances (tokens) of pain are identical with actual neural events. On a token-identity version of psycho-neural identity, my particular headache today might be identical with a certain instance of C-fibers firing, even if your particular headache yesterday was identical with a given neurotransmitter process. There is nothing to an actual experience of pain over and above the token brain event with which it is identical.

Retreating to token identity lost the theory its credibility. Token identity could not explain why instances of the same type of pain, say headaches, might be identical with C-fibers firing in one case but neurotransmitters in another. The alleged identity became even more questionable if those correlations (my pain with C-fibers firing; yours with a neurotransmitter release) were not regular and lawlike.

To make matters worse, empirical results coming out of neuroscience at the time raised additional doubts about universal *unipotentiality*, the one-to-one identity between one brain and one mental event on which the identity was based. Brain mapping of amputees revealed that the region of the brain usually associated with the amputated limb could be coopted by a different function. Areas in the brain corresponding to one faculty can be taken over by another faculty if necessary and put to a different use. Regions of the brain dedicated to vision could be taken over by hearing or touch, for example. The identity theory could not account for such widespread plasticity.

Although some localization was discovered, clinical research soon focused on this dramatic plasticity, its *pluripotentiality*, and its often distributed architecture. Multiple realizability also appeared to be a common characteristic of biological organization. Certain neurotransmitter molecules have a particular effect in the digestive tract but an entirely different effect in the brain. Also, complex neural networks were discovered to be widely distributed throughout the brain.[2] Thoughts and beliefs—mental events generally—could not be easily and unequivocally identified one to one with local physical structures or processes.

Combined, these observations prompted the fundamental question, "What justifies identifying token mental events with a diverse array of brain events that seem to share no common or general feature with each other (other than being brain events)?"[3] Since neither brain imaging nor other probes at the time had sufficient resolution to identify microneural processes exactly, both versions of the identity theory were left with the following unpalatable choice: either a *double aspect theory*, for which mental events are ontologically identical to physical events, but physical stuff has two qualitatively distinct aspects, physical and mental,[4] or what might be called *promissory physicalism*, the pledge that once brain imaging techniques improved, the exact correlation between thoughts, intentions, or sensations and neural patterns would be elucidated. In recent years, the explosive power and granularity of brain mapping technology continued to up the ante on that pledge (but see Mitchell 2021).

And so, the themes presented in our introductory chapter reappear in the twentieth century in the guise of the philosophy of mind's theories about *type identity* and *token individuation* (the identity theory called it token identity theory, not token individuation theory). What makes a particular event or process a token of a given type? And how can actual instances or specimens become increasingly individuated and therefore uniquely themselves, all the while continuing to belong to the same type? What constitutes and holds together a type of mental event like an intention, and how is it biologically realized, one to one? How is that even possible in cases of intentions like running for office, which take years if not decades to be realized? The standard assumption that universal and unchanging primary properties confer type identity makes this question a critical one. Lacking an account of coherence that binds distinct individual and previously separate neurons into a unitary network with emergent properties while simultaneously allowing that network to individuate and differentiate over time as distinct tokens, both versions of the identity theory failed to account for emergent properties of interdependencies even among physical (neural) processes.

Like behaviorism before it, the token identity theory also egregiously failed to explain the emergent content of thoughts and intentions and the qualitative feel of sensations like a searing pain, a vivid color, or a haunting fragrance. As noted earlier, the subjectively felt properties of the latter were labeled qualia. Even if one-to-one correlation was established between actual neural processes and particular qualitative feels such as the experience of seeing the vivid crimson of Don Juan roses, correlation alone would not explain what it is about those neural events that produces that qualitative experience of color—or of a searing pain. It would fail to account for qualia. As formulated in a commonly used example, a blind neuroscientist who knew everything about the physical substrate that correlates with the color red would still not fully grasp everything about the experience of perceiving a Don Juan rose. Finally, the purported identity between neural processes and mental events also failed egregiously to account for subjective experience (the hard problem of conscious awareness in general and self-consciousness in particular). Neither version of the identity theory could *explain* why certain neural patterns turned on the light of awareness nor account for how brain events can be about facts in the world. Much less could it explain recursive self-awareness.

Functionalism

Inspired by the work of British mathematician Alan Turing on universal computing machines (Turing 1936), the burgeoning fields of computational devices and information theory transformed Anglo American philosophy of mind. Specifically, the fact that devices can carry out the same information processing on different hardware suggested to philosophers that the multiple realizability of mental events might be reconceptualized on the analogy between a computer's hardware and its software—in particular, on the fact that software can be instantiated in a variety of hardware. This theory came to be known as *functionalism*.

According to functionalism, what makes an experience a mental state is not its internal and material substrate (its primary properties) but its function. On the analogy of software, function is realized as a series of if–then relations (algorithms) that link physical inputs to outputs such as behavior. Functionalism thus updated behaviorism's ideas about conditioning with conceptual tools borrowed from information-processing: mental events just are the algorithms running on a brain.

By decoupling the informational and functional aspect of mental events from its material realization—in transistors, solid state chips, or organic matter—functionalism decoupled feelings, emotions, thoughts and intentions—mental processing generally—from biological wetware. Whereas the identity theory's emphasis on organic stuff had left no room for artificial intelligence, functionalism also pleased science fiction buffs: if the same software happens to be running on silicon chips instead of organic matter, we can conclude that the silicon device feels pain. Artificial Intelligence as a serious field of research was born.

Claiming that the functional order and organization according to which actions are governed is more significant than the stuff the organization arranges conferred an unavoidably dualist tint to the theory. As a result, functionalism came to be referred to as *property dualism* by advocates and detractors alike. It is noteworthy for our purposes, however, that one early critique of functionalism pointed out that it made no room for context (Dreyfus 1972).

Chinese Room Objection

The idea that input–output relations can fully account for all higher-level mental capacities, including emergent properties such as qualia, consciousness, understanding, and so on, was soon challenged. The so-called Chinese room argument of American philosopher John Searle (Searle 1980) is often considered the metaphorical coup de grâce to functionalism. Suppose you are locked in a room, and a sheet of paper with a question written in Chinese characters is passed to you under the door. By consulting a Mandarin-English dictionary (which serves as an algorithm-like lookup table), you respond with other characters. Searle's point is that there is no reason to suppose that by enacting what is in effect a software program in this manner you understand the meanings of the interlocutor's squiggles, or that your own squiggles convey an intended reply.

In short, functionalism's failure to address understanding, meaning, the intensional and intentional content of mental events, qualia, and the hard problem of consciousness was as egregious as its predecessors', the identity theory and behaviorism.

Intentional content with a *t* is what a mental event such as a thought or belief is about. But what is inten*s*ionality with an *s* with respect to mental events? In addition to qualia, among the novel properties of mental events is *intensionality*. As noted in the previous chapter, intensional concepts and properties like *here* and *now* are those whose character is inherently context dependent. As an example of intensionality in belief statements, consider the following: Ann believes that John's wife is cheating on him. Ann does not know that Mary is John's wife. Under these circumstances, the term *Mary* cannot be substituted for *John's wife* in the statement "Ann believes that John's wife is cheating on him"—because Ann does not believe that it is Mary who is cheating on John.

And yet intensional content of mental events, such as Ann's belief above, is at the heart of how intentions cause actions. Add a premise to the above case: suppose Ann believes that John is also cheating on his wife, and Ann tells Mary that. Since Ann did not intend to tell John's wife that he is cheating on her, the two terms cannot be substituted salva veritate in these sentences either. Ann's comment was informed by the intensional content of her beliefs, which, in this case, do not link Mary with John's wife.

The implications for functionalism were clear: any functionalist account of intentional action must account for how purposive behavior is directed by mental content as conceived by the agent. Ann's beliefs and intentions, and her comments, are conditional on her mindset, that is, on her conceiving of the person as John's wife, not as Mary. Despite referring to the same individual, the terms cannot be substituted for one another without loss of truth value in sentences where these appear as the intensional content of mental events.

Intensional content of mental events must be indexically defined because beliefs, thoughts, and intentions as generated by contextual constraints are perspectival. They are products of context-dependent constraints. Any account of *intentional* causation must therefore account for the role of *intensional* content in purposive action.

Principle of Supervenience

Aiming for a *nonreductive physicalism* that could account for emergent mental properties like meaningful and intensional content as well as qualia, the Principle of Supervenience was formulated to buttress functionalism and, in passing, further refute the identity theory.

Advocates of the supervenience thesis acknowledged that mental events have remainder properties that do not match neuronal patterns defined extensionally. The intensional content of thoughts, beliefs, and the felt experience of color and pain cannot therefore be identical to either neuronal events or software packages.

Advocates of the new theory proposed instead that mental events supervene on brain events.

The *supervenience thesis* claimed to offer a nonreductive yet fully naturalist approach that avoids the pitfalls of both the prevalent psychophysical identity theory and Cartesian dualism. Korean American philosopher Jaegwon Kim, who had earlier espoused the identity theory, was its best-known advocate. Because of the subjective, qualitative character of some mental events as well as their intensional content, mental events could no longer be considered identical to either dematerialized syntactical relations or neural processes tout court. Kim proposed instead that mental events are in an asymmetric relation of supervenience with brain events.

As American philosopher Donald Davidson first used the term in 1970, the supervenience thesis held that when one type of properties (Ψ)—those that define mental events—supervenes on another type of events and properties (Φ)—physical ones—there can be no changes in Ψ without a corresponding change in Φ. No two events can be alike in all underlying or subvening (physical) properties while at the same time differing in their emergent or supervening (mental) properties. An event "cannot change in some mental respects without a corresponding alteration in some physical respects." Resolutely naturalist, the supervenience thesis held that mental processes are not identical to but depend on physical ones.

Davidson argued, however, that on pain of turning into another version of the identity theory, that dependence relation must be causal. Subvening neurological events must actively cause supervening mental properties (Davidson 1980). Suddenly, the problems of mereology reappeared. How do different subvening physical (brain) processes cause supervening thoughts and beliefs with intensional content, for example (Juarrero 1999)? And especially, what role does the intensional content of those mental events play with respect to causing purposive action? Can intentions cause actions in virtue of their supervening properties, their intensional content in this case? The problem of top-down causation thus returned as well.

Recall that the received understanding of causal relations to this day maintains that causal power goes only one way, bottom-up. Davidson and Kim took for granted that proper causes only operate as efficient causes, that cause and effect must be spatiotemporally distinct, and that causal power goes only from subvening properties to supervening ones. As merely aggregative sums of subvening brain events, moreover, mental events must be causally powerless because if the emergent properties of mental events as mental could bring about changes in the physical world, this top-down causal influence would violate physical closure and overdetermine the universe.

The only acceptable conclusion for the supervenience thesis was that the supervening properties of mental events might be real, but they are epiphenomenal, that is, impotent.

In a scathing critique, Kim (1989, 1998) labeled this crushing implication of the supervenience thesis "Descartes' Revenge" because of its analogy with the fatal flaw of Descartes' dualism described at the beginning of this book. Supervenience's dirty little secret was that avoiding overdetermination and violating causal closure unavoidably commits its advocates to epiphenomenalism and the disavowal of top-down causality. Supervenience cannot explain acting purposively—*because* of the context-dependent intensional content (much less qualia) of an intention—without committing the same errors as substance dualism. Qualia and the intensional content of intentions can only be the impotent effluvia of physical—in this case neuronal—processes.

In the end, Kim concluded that, pace purported claims of supervenience, mind and body are in fact identical after all (despite their distinct properties). To accentuate the pertinence of the supervenience debate for this work, I reiterate that the impasse between emergent properties and causality arises in the first place because philosophy has no general understanding either of the origin of mereological coherence or of top-down causation from wholes to parts, or other than in terms of efficient causes.

Multiple Realizability

The plausibility of both functionalism and the supervenience thesis vis-à-vis the token identity theory nevertheless rested as strongly on the fact of degeneracy as did the initial attempt to formulate the identity theory in terms of types, not tokens. Understanding coordination dynamics and hierarchies as constraint regimes generated by their constituents, each multiply realizable by the relata that comprise them is key to rethinking these problems.[5] Multiple realization is not limited to neurological and information-processing dynamics. As we have seen throughout this work, it is not even limited to living things, where it is called degeneracy.

Multiple realizability appears to increase throughout the universe alongside complexification. In a wide variety of fields in addition to the computational and cognitive neurosciences, including empirical neuroscience, we have presented evidence that emergent properties such as autocatalytic cycles and convection cells can be variously realized. Phenotypes can vary, sometimes quite dramatically, depending on the context in which gene expression occurs. This provides evidence that realizations of a genome's information content in a particular phenotype are not one to one. This is also clearly so with respect to the intensional content of beliefs and intentions. Even assuming schizophrenia has a strongly genetic component, for example, paranoid schizophrenic patients in medieval times were not fearful of the CIA. The role of context in multiple realizability and intensional content must be accounted for. Furthermore, higher-level (supervening) properties can persist despite dramatic changes in the underlying neurological realizations. Different neural patterns can realize the same thought; the same neural pattern can subtend different emotions—depending on context. As noted earlier, persistence is grounded in the lack of one-to-one correspondence between changes in lower-level events and particular functions or properties. Plasticity, degeneracy, and pluripotency are typical of the relations between types and tokens.

Remarkably, however, the most trenchant objections to the supervenience hypothesis initially challenged the claim that mental properties are degenerate at all (that is, multiply realizable). Of those questioning the multiple realizability premise itself, American philosophers Thomas Polger and Lawrence Shapiro (Polger and Shapiro 2016) are among the best known. Those arguments are discussed in detail in the next chapter, where the tendency to question anything but the underlying framework about causality and the reality of mereological relations is highlighted.

Before that, however, let us close this chapter by bringing the backstory up to date with the 4E approach.

The 4E Approach Today

In 1975, Hilary Putnam published "The Meaning of 'Meaning,'" where he argued that "the mind ain't just in the head"; it is extended and distributed throughout the body. This take on the mind and cognition contrasted vividly at the time with the approach that held that the mind is entirely located inside the brain, where it mysteriously processes contentful representations and produces intentional actions.

Clark and Chalmers's (1998) article directly addressed the question, "Where does the mind stop and the rest of the world begin?" by introducing the philosophical thesis of *active externalism*: it is not that the mind ain't just in the head and extends to the rest of the body. In fact, no principled separation can be drawn between mind, body, and environment. Human activities incorporate objects in the environment into cognitive and affective processes; these, in turn, extrude into the devices and tools they design and deploy.

Earlier, *Origins of the Modern Mind* (Donald 1991) had presented evidence that cognition and consciousness actively involve tool and technology use. From paper and pencil to handheld calculators, such devices at first glance seem only to store and execute the products of the mind. In fact, Donald argued, a wide range of tools must be considered co-creators and co-implementers of mental activity. Our mind extrudes into the world through its embodiment in a panoply of devices; conversely, those devices become integrated into our cognitive and affective framework. It is evident that this idea is congruent with the one presented in this book.

In the past twenty-plus years, this approach has grown into a movement. Often labeled the *4E approach*, it draws from complexity science and dynamical systems theory to describe cognition not only as embodied throughout the human anatomy and physiology (Thelen and Smith 1994) and extended into technology, tools, and other devices (Chemero 2009). Ontically, it defines mental events as enactments of social practices and activities (De Jaegher and Di Paolo 2007, Noe 2010).

Sensorimotor Life (Di Paolo, Buhrmann, and Barandiaran 2017) explicitly refers to recursivity, reciprocal causality, and dynamics as central to the enacted mind. The authors note that human activities are "highly context-sensitive and full of constraints of all kinds, including, but not limited to time pressures at multiple scales, context-sensitive manipulation of tools and artifacts, the need to coordinate actions with variable objects and other people . . . constraints on the capabilities and demands of the body" (Di Paolo, Buhrmann, and Barandiaran 2017, 4). Slaby and Gallagher (2014), too, extend the idea of enacted minds to institutions and organizations. They consider science, for example, to be a cognitive institution that "shapes our cognitive activity so as to constitute a certain type of knowledge, packaged with relevant skills and techniques" (33). One can consider the elaborate social practices of cassava and nardoo preparation to be instances of enactivist practices.

Radical enactivism goes even further: it posits that mental, cognitive, and affective processes are constituted in and through sense-making activities. This version of the 4E approach conceptualizes cognitive and sensory processes as radically context dependent on activities: cognitive processes are nothing but the enactment of social practices. Specifically, this radical version casts doubt on the reality of mental representation and intensional content as such. From this perspective, the human self (Kyselo 2014) in its entirety is also constituted as (nothing but) a "social existence" organized as self-other activities "represented by a phase space . . . whose attractors can be defined" (Kyselo and Tschacher 2014, 1).

There is significant agreement between the ideas developed here and 4E's insight that embedded and embodied human practices and activities provide the right framework for understanding minds and mental processes. However, although the role of constraint in generating and preserving structured self-other possibility spaces (such as sensorimotor habits, social practices, and schemes) is implied in most of the 4E publications, when constraint is mentioned at all, it is used almost exclusively in the stabilizing and restricting sense of constraint.

In contrast, this book has attempted to articulate the manner in which enabling constraints generate those very self-other interdependencies and practices that are at the heart of the 4E approach. Context-dependent constraints create the phase space, attractors, coherences, and mutual dependencies that the 4E approach emphasizes. Constraints are also the source of the emergent properties of those interdependencies, among which, this book has maintained, are intensional content, qualia, and phenomenal awareness, which in turn and acting as governing constraints produce, regulate, and modulate socially required, task-appropriate, and context-sensitive behaviors.

This book has also attempted to explicitly rethink the "[recursive] causality unknown to us" that Kant argued was at the heart of purposive behavior by proposing a general framework of constraints for mereological relations. It has delved deeply into how constraints facilitate and impede information flows between those layers of natural and social hierarchical organization we call individuals, their habitats and worlds, and the social practices in which agents are embedded and which they co-create in their enactments. I have done so by returning to the source of my original interest in complexity theory: the parallels between Kant's understanding of intrinsic teleology as self-organization and the recursive causality at work in Prigogine's dissipative structures (Juarrero-Roque 1985).

Approached with the passkey of constraints and insights from biologically inspired hierarchy theory, far-from-equilibrium thermodynamics show that constraints are not efficient causes; they do not transfer energy directly and are therefore not vulnerable to charges of overdetermination or violation of causal closure. They are coherence makers and sustainers. Body–social interdependencies, agent–environment couplings, patterns of sensorimotor schemes and habits, and the array of interdependent activities at the center of the 4E approach are therefore best understood as the outcomes of bottom-up enabling and generative constraints. They embody and enact new coherences and as such drive major transformations in evolution. In both the living and the nonliving worlds, constraints generate coordination dynamics with emergent properties. Once coalesced and acting as top-down constitutive/governing constraints, the order parameters of those extended coordination dynamics stabilize, preserve, and realize those interdependencies in the enactment of a variety of activities and practices. In cascades of negative feedback processes thanks to the nonlinearities that characterize far from equilibrium dynamical systems, governing constraints regulate and modulate energetic processes such that individuals can enact those social constraints as socially sanctioned activities and practices.

Without the role of constraints in generating those multiply realizable interdependencies, the self-sustaining mereological relations of "sensorimotor schemes and habits," for example (Di Paolo, Buhrmann, and Barandiaran 2014), would remain unexplained. Accounting for how constraints take conditions away from equilibrium and/or away from independence gathers habits, repetitive drills, sedimentation, self-sustaining interdependencies, and the rest of the conceptual tools of the 4E approach under the general framework of constraints. Without a role for constraints, phase transitions to new possibility spaces with emergent properties would remain unaccounted for. Finally, since as noneconomic entities constraints do not directly transfer energy, it is the conservation of constraints that accounts for the covarying behavior patterns and sensorimotor coordination on which the 4E approach is centered.

In short, *Context Changes Everything* has aimed to explain the sources of the 4Es, embeddedness, embodiment, enaction, and ecological coherence and interdependence. It did so by examining the manner of causality whereby coupling and linkages that generate those 4Es come into being, persist, and even become sedimented and entrenched.

The claims made in this book are therefore more far reaching than those of the 4E approach. Rather than address the topic of mental representation directly, *Context Changes Everything* focused on the role of emergent properties more generally. It proposed that enabling and/or constitutive/governing constraints are in place not only in living things capable of sensorimotor schemes. Physicochemical convection cells, planetary atmospherics, lasers, and even homeostasis in eukaryotic cells are only a few examples of precursors mentioned in this work. Even prior to the emergence of aquatic life and land-based lichens and mycorrhiza, these constraints continued to complexify in hydrothermal deep-sea ocean vents. Each of these these transitions describes coordination dynamics that are the products of context-independent and context-dependent constraints. A continuum of constraints has evolved, beginning in the Big Bang and the creation of elementary particles, through galactic formations and planetary atmospherics, to the emergence of increasingly symbiotic life and conscious awareness, and finally to the myriad social practices and activities that characterize the Anthropocene and which the 4E authors highlight. Well before the origin of life, that is, constraints at work in the abiotic domain generated and governed precursors of the human self-social dynamics described as the 4Es.

Since embeddedness, sensorimotor habits and schemes, and other 4E concepts presuppose the existence of mereological relations of constraint between parts and wholes, let us now examine more closely the role of multiple realizability itself in bringing about functional dependencies.

14

Multiple Realization and Supervenience

A Philosophical Case Study about Constraints

This chapter analyzes two contemporary arguments about multiple realizability and the relation between supervening and subvening entities and processes. Lawrence Shapiro and Thomas Polger's views (Shapiro 2000; Polger and Shapiro 2016) on functional multiple realizability are discussed first. I then examine Marc Lange's (2017) logic of explanation as it pertains to natural phenomena that are alike.

* * *

The Principle of Supervenience turned on the relation between subvening processes (also called realizers) and supervening (realized) properties generally, not only in the mind–body realm. Its claims presupposed that the following questions had been answered in the affirmative: (1) Does multiple realization actually happen? (2) If so, do multiply realized (supervening) types identify real kinds? What qualifies as different tokens of the same kind? For example, Are camera eyes, like those of humans and octopi, and compound eyes, like those of insects, different realizations of the same kind of thing, eyes? Or are they two different kinds of things?

As we saw with the two versions of the identity theory, the classical question of type identity—are types real or not?—has reappeared. Or does reality, and, especially, causal power inhere only in ground-level primary properties, making types mere epistemic classifications?

Shapiro and Gillett on Multiple Realizability

American philosopher Lawrence Shapiro reaffirms Davidson's premise that only causally relevant properties, those that cause a thing or process to carry out a function, qualify as realizing properties. Kinds are defined in terms of type-level properties. In the case of mousetraps and corkscrews, those properties are functional: this device is a mousetrap if it performs the function of catching mice; that levered gizmo is a corkscrew because it carries out the function of uncorking bottles. It is only in virtue of those causal powers that they are the type of thing they are, corkscrews or mousetraps.

Shapiro maintains that two entities are distinct token realizations of the same kind only if their causal mechanisms are different. Even intuitively, differently colored corkscrews do not count as different types of corkscrews. Tokens must realize the same functional type through different causal mechanisms. Silicon and biological brains, for example, differ in material composition (one is made of organic neurons, the other of silicon wafers). But since electricity is what causes information processing in each, "There seems to be no more reason to count [silicon and biological brains] as distinct realizations" of mental events (Shapiro 2000, 645) than if some neurons were stained gray and others purple.[1]

Shapiro justifies this conclusion by referring to Fodor's principle that "what justifies a taxonomy, what makes a kind 'natural,' is the power and generality of the theories that are enabled when we taxonomize in that way" (Fodor 1981). Scientific theories discover similarities that enact lawful regularities; it is the scope and stringency of the natural laws they subtend, not mere correlation, that underwrite the power and generality of truly scientific predictions. So, to qualify as multiple realizability, causal mechanisms of allegedly distinct realizations must "require different manipulations, *[and be] described by different laws*" (Shapiro 2000; emphasis added).

From this premise, Shapiro concludes that functional properties are either not ("all that") multiply realizable, or there is nothing particularly interesting about higher-level functional phenomena as such.

He reasons as follows: mammalian and octopus eyes share the same causal mechanism: light impinges on a single lens, projecting an inverted image onto a retina, which converts the image into an electrical signal. Because the single lens, retina, and electrical signaling mechanism is the same in both mammals and octopi, there are no causal differences in virtue of which octopus eyes and mammal eyes realize vision. There is a fortiori no difference in the natural laws that bring about vision in the two cases. Since a "virtually identical structure" (Putnam 1967, 1975) governed by the same natural law causes vision in each[2] (Shapiro 2000; Polger and Shapiro 2016, chap. 3), mammal and octopus eyes are not multiple realizations of the same kind of thing, eyes.

If a predicate does not feature in a natural law, it does not pick out a real kind and so no multiple realization occurs. In philosophical jargon, only *projectible properties* appear in natural laws. By identifying traits governed by natural laws, projectible properties underwrite prediction; as such they support counterfactuals and can "be taken as guides to valid inductions."[3] That is what makes them real properties.

Shapiro argues that functional properties such as uncorking wine bottles and catching mice are not projectible because there are no natural laws that determine uncorking or mouse-catching. He goes further. The "sciences" of economics, ecology, and so on are also bogus, and for the same reason: they deal in bogus kinds because no predicates in those sciences are projectible. There are no natural laws of economics or ecology. Consequently, there are no proper natural laws in the special sciences. Indeed, there are no natural laws that hold for higher-level or purportedly functional properties as such.

The unavoidable implication for philosophy of mind is that thinking, intending, sensing, perceiving, and other higher-order (mental) events are projectible traits only if those characteristics feature in natural laws. But no psychobehavioral laws exist that universally and deterministically link mental properties as such with particular actions. So mental properties do not pick out real ontic kinds. They are not real properties. "If two realizations contribute to a capacity in causally distinct ways then this must mean that there are no *[lower-level]* laws common to both of them" (Shapiro 2000, 648; emphasis added); so each realization enacts a distinct natural law. Different corkscrews, mousetraps—or mental and social processes —might be functional isomorphs of one another, but they are not multiple realizations of real kinds.

American philosopher Carl Gillett (2003) describes this impasse as the clash of two opposing views on causal relations between realizer and realized. Under Shapiro's *flat realization view*, as Gillett terms it, the fact that the lenses of mammals and octopi eyes are composed of different proteins and rely on different pigments is irrelevant. These features have no direct causal role in bringing about vision. The real causal mechanisms are the single lens, inverted image, retina, and electrical impulses, which mammals and octopi share. Shapiro (2000): "Sameness in the processes that result in the formation and analysis of an image screen" provides truth conditions for determining causal relevance. Differences in proteins and pigments must be screened off as causally irrelevant to the actual and virtually identical causal mechanism in both.

The flat view, that is, requires that properties and causal mechanisms of subvening realizers directly cause the realized and supervening properties. The combination of realizing components and processes (here a single lens projecting an inverted image on a retina, which converts it to electrical signals) is what causes vision. The flat view, in short, implies a bottom-up and strictly determinist one-to-one correlation between realizer and realized (which, as efficient causes, must also be separately identifiable). Because natural laws project microlevel properties, only one macrostate realization results from a given set of microstates.

In contrast, the *dimensioned realization view* has a more expansive interpretation of cause: on that view, the powers of the realized (functional) property result from the "powers of the realizers" (Gillett 2003, 595) even if the realizers "may and often do contribute no common powers including those individuative of the realized property" (Gillett 2003, 602). As an example, individual neurons do not think or feel, yet thinking and feeling result from systemwide emergent properties of collective neuronal processes. The dimensioned realization view, in short, recognizes that, together, realizers can induce emergent qualitatively novel properties. Gillett (2016) notes that whereas physicists have long accepted that collective properties of nesting entities manifest emergent properties beyond those of their nested components, physicists still fail to recognize the active causal power of collective properties to change the go of events.

Unlike Shapiro, therefore, proponents of the dimensioned realization view take octopus and mammal eyes to qualify as different realizations of the same kind of thing. American philosophers Jerry Fodor and Ned Block argue that differences in pigments and proteins of lenses in mammals and octopi make those two instances multiple realizations of a distinct type of entity, eyes. "What is important is that the properties/relations of different realizations, such as lenses composed of different pigments and proteins, nevertheless contribute powers that result in the powers individuative of having an eye" (Gillett 2003, 596). Unlike corkscrew color, which contributes nothing to the device's uncorking powers, differences in proteins and pigments in the eyes of mammals and octopi do contribute to realizing vision in the two cases.

Significantly for our purposes, Gillett also points out that under the dimensioned view, although realizers are not technically causes because they are not spatiotemporally distinct from the realized properties, components can contribute powers that result in but do not literally (efficiently) cause the emergent properties of the whole. On the dimensioned realization view, constituent parts and processes are labeled potential realizers, not causes, of the higher-level property (Gillett 2003, 596, 601). The common set of powers of the realized property (Figdor 2010), goes beyond the powers of the realizers separately.

This contemporary disagreement about mereological causation revisits the entire history of the subject matter; the dimensioned view also makes room for the idea that enabling constraints generate multiply realizable supervening types with emergent properties (vision, in this example). Realizers of the same higher-level property can differ at the microlevel, depending on context. Nevertheless, they remain variant tokens of the same, type-defined, emergent constraint regime. The deep dyslexia example presented a dimensioned realization view of meaningful causation.

Throughout the debate, Fodor, Block, Shapiro, and other contemporary philosophers who study multiple realizability do not seriously consider the possibility that a form of cause other than efficient cause might, mereologically, bring about emergent, higher-level (realized or supervenient) properties. Shapiro explicitly maintains that only lower-level properties ground natural laws and guarantee their truth. He adheres to the received view that the explanatory arrow always points downward. From his perspective, only bottom-up causal relations underpin scientific explanations; only microlevel properties and causal mechanisms are projectible and support universal and determinate natural laws that hold across instances. It is only because identical causal mechanisms bring about vision in mammals and cephalopods (lenses, retinas, etc.) that the science of vision can study "optical principles that apply to light and lenses"[4] generally, he maintains. These correlations hold true "in virtue of some deeper set of laws" (Shapiro 2000, 653).

Having concluded that no higher-level functional properties are projectible or ground universal natural laws, Shapiro concedes, "There may be general statements about the higher levels that refer only to their capacities" (Shapiro 2000, 648), such as statements like "Lever corkscrews and rack-and-pinion corkscrews are both corkscrews," and "Natural selection selects organisms that can maintain their body temperature over those that cannot." Such propositions, however, are analytic and tautological, not proper natural laws. Their impoverished content is why most scientists aim to causally connect common lower-level internal mechanisms, processes, and properties with supervening properties and functionality, he maintains. Proper sciences thus aim to discover laws that correlate internal mechanisms with supervening properties (Shapiro 2000; Polger and Shapiro 2016). Only fundamental properties cause supervening functions, and functions are the direct effect only of relevant lower-level causal processes. Degeneracy or multiple realizability are therefore bogus ideas (See Figdor 2010 critique). All the heavy causal lifting is done by the subvening processes.

Gillett's 2016 book, in contrast, is a plea for *mutualism,* the view that emergent properties even physicists if not philosophers recognize to be real are also not epiphenomenal (which most physicists still espouse). But Gillett himself does not address the manner of causality in virtue of which the emergent powers of what he calls collectives and system-wide interdependencies might exercise that power on components and in behavior.

Lange—Because without Cause

Can Shapiro and Gillett's positions be reconciled? Contemporary American philosopher Marc Lange (2007, 2017) argues that not all scientific explanations appeal to causal force laws. Questions of the form, "Why are gravitational and electrical interactions alike in conserving energy?"—or more generally, why are X and Y alike with respect to Z?—cannot be explained solely by deriving gravitational interactions from gravitational laws, and electrical interactions from laws governing electrical interactions. Individually or jointly, such force explanations make no mention of why the two cases would be alike with respect to energy conservation.

The point is a general one: "Causal explanations do not unify two cases" from different fields (Lange 2017, 57). Neither do compound explanations such as combining appeals to gravitational and electric laws explain why two actual instances of gravity and electricity are alike with respect to energy conservation.[5] In particular, separate and joint explanations that make reference to forces cannot rule out the possibility that the two instances conserve energy just as a mere coincidence.

Whereas Polger and Shapiro dismiss appeals to noncausal relations, Lange argues that some scientific explanations do explain *alikeness* by reference to shared higher-level constraints and more inclusive dimensions (Lange 2017), not common efficient causes. Not only can shared constraints in a more encompassing dimension explain why X and Y hold independently (Lange 2017, 65), appeals to constraints in common can also explain why the individual force laws hold in the two cases (Lange 2017, 61). Let us see why.

Principles and Laws

As noted, in the natural sciences the term *law* is commonly used to designate only those correlations that support counterfactuals; natural laws commonly identify force laws underwritten by projectible predicates. Natural laws so understood ground the "covering-law model of explanation" that contends that events are successfully explained if and only if they can be inferred from a natural law together with initial condition statements. It is for that reason that philosophers like Shapiro refuse to apply the labels *lawful* and *scientific* to the special sciences; they are dismissed because their purported laws lack the necessity and universal scope to support inferential predictions.

Lange contends that even in those cases where entailment holds, proofs deriving the explanandum from force laws would still not show why two cases are alike with respect to a certain property despite being the outcome of different forces. Satisfactory explanations of the relation between the two cases must show not only why gravitational and electrical interactions are alike in conserving energy, they must also show why their joint occurrence is not coincidental. Coincidences, to put it crudely, do not have a "common reason" (Lange 2017, 66) for happening. (Note use of *reason* in lieu of *cause*.) In contrast, specifying a real and "distinct class" of which the two cases in question are instances would identify a common reason for why the two cases are alike. By identifying that they are similar "in virtue of some context where the results exhibit this noteworthy similarity" (Lange 2017, 280), explanations of this form would succeed in showing why distinct cases must be alike.

Lange emphasizes this is not merely an epistemic move. Epistemic and psychological considerations alone are insufficient to reveal what makes certain combinations of facts explanatory and not merely coincidental (Lange 2017, 287). Purported explanations are satisfactory depending on their ontic implications. In this example, reference to "energy conservation" moves the explanation to a shared ontic context in virtue of which the similarity is necessary. In mathematics, finding a shared "class of cases" might require moving to a third dimension and showing that the lower-dimensional outputs are alike by virtue of being instances of a class defined in that higher dimension—by being realizers of a common constraint, that is. "Metaphysics must not foreclose [on why-type] explanations . . . on pain of failing to do justice to the fact that science has rightly taken such proposal seriously" (Lange 2017, 67).

This requires taking relations and context seriously. This book has claimed that enabling and governing constraint regimes whose interdependencies weave together a common context within which individual instances are distinct realizations provide that why-type explanation. Constraints, in short, provide the reason why distinctly realized entities are tokens of the same kind or type of thing.

* * *

Lange notes a second difference between shared constraints operating in a third dimension, on the one hand, and force explanations, on the other: constraining principles that apply to both domains explain the (force) laws in question by describing a common context that excludes some alternatives while including others in its common possibility space. The coordinates and boundaries of that shared context limn what is empirically possible and what is not.

Possible force laws are force laws that satisfy constraints laid down in that third dimension. *Logically possible force laws* that cannot empirically satisfy those constraints are *factually impossible* (Lange 2017, 51). By underwriting "a certain distinctive kind of invariance under perturbations" (Lange 2017, 48), conservation principles—the shared dimension—serve as context-independent constraints on a shared space of *empirically possible force laws*, be they gravitational or electric.

That is, shared constraints from a common dimension define a more expansive possibility space where distinct but logically possible force laws can operate simultaneously. Conservation and symmetry principles—common dimensions of constraint in physics—do just that; they set the boundaries or coordinates of what is simultaneously realizable empirically and what is not. They determine context-independent constraints. The International Monetary Fund does the same; it does not lend much money directly, nor does it negotiate with a country and its creditors. The IMF "draws the boundaries of possibility and policy" (Lustgarten 2022) within which other banks, investors, and rating agencies must operate. It does so by setting the common boundary constraints that would govern any natural realization.[6] Lange's third dimension thus corresponds to a shared background of context-independent constraints. This shared possibility space outlines the context in which invariances can hold and persist.

Lange prioritizes natural over logical possibility and impossibility. He cashes out natural necessity and impossibility in terms of invariance: conservation and symmetry principles establish "a certain distinctive kind of invariance under counterfactual perturbations" (Lange 2017, 48). Conservation and symmetry "would still have been conserved even if the forces at work were different" (Lange 2017, 48), "even if there had been additional *kinds* of forces threatening to undermine its conservation" (Lange 2017, 72; emphasis added). Because conservation and symmetry principles are multiply realizable, they can account for laws that could logically be in effect (but factually are not).

In such cases, the arrow of explanation of necessity points upward to a common dimension of constraint.

* * *

How stringent must the invariance preserved by shared constraints be to qualify as scientific? Lange's case study is about reconciling instances of electrical and gravitational interactions by reference to energy conservation. In his example, the common dimension refers to "great general principles which all [force laws] follow" (Lange 2017, 51). Principles of energy conservation and symmetry are indeed universal; they "sweep across"[7] all other laws. By grounding "alikeness" in a universal and an unchanging dimension, they account for "why the [two cases] are alike in possessing certain features" (Lange 2017, 51). Despite being realized by different force laws (gravity and electromagnetism), *energy conserving* is a projectible predicate; it features in a universal principle, a shared and universal constraint regime.

In contrast, our own examples of constraints such as feedback and catalysts do not universally "sweep across" conditions universally. Their context dependence prevents them from grounding counterfactuals universally.[8] Is this a fatal objection to this book's project? Association and correlation are not causation; something stronger is required to justify a "causal" claim. We have suggested that, within a specified context, constraint regimes of mutual dependencies effect covariance and thereby ensure an interactional type's persistence. But must the counterfactuals in question hold universally?

I reject this assumption and the requirement of universally projectible predicates. Among the central claims of this book has been that context dependence is a feature (not a bug) that science (Science 2.0?) must incorporate into its conceptual framework. I reject the unstated premise that ontic kinds (which we have called interactional types) are solely underwritten by laws and principles that apply in all instances, anywhere and always. Requiring that all real events and processes refer to universal laws that support prediction under all conditions rules out, from the outset, any effective role for context dependence. That is the framework of modern science and philosophy. This approach was already tried with Platonic Forms, Aristotelian and Cartesian substances, and Laplacian laws that are allegedly unchanging and eternal regardless of context. It is a framework that works spectacularly well in many cases, such as with predicting the motion of two planetary bodies, for example. But there are cases, such as whirlpools and coupled pendulums, not to mention ecology, psychology, and other special sciences such as economics where it does not. In those fields, context changes everything.

In this connection, and in opposition to a Laplacian notion of natural law (equations that describe events at the most microscopic and fundamental level, from which all macrophenomena are held to follow and from which they can be deduced), Herbert Simon notes that a so-to-speak Mendelian notion of natural law

> takes as its ideal the formulation of laws that express the *invariant relations between successive levels of hierarchic structure*. It aims at discovering as many *bodies of scientific law* as there are *pairs of successive levels*. . . . . The fact that nature is [contextual, path dependent, nested, or] hierarchic doesn't mean that phenomena at several levels—even in the Mendelian view—cannot have common mechanisms. (Simon, in Pattee 1971, 24–25; emphasis added)

Or, on a generous understanding of "mechanisms," common constraints and shared contexts.

In the spirit of Simon's comments, this book calls for expanding the framework of the natural sciences to include what might be called *effective science*, where alikeness, lawfulness, and counterfactuals identify invariances conditioned on multiple shared constraints in specified possibility spaces. Complex dynamical systems theory calls for a science that includes indexicals and their contextually situated properties.

Effective science would seek counterfactuals that hold conditional on a specified range of contexts: this focal heterarchical level, in this embedding heterarchy, and realizable by a specified range of tokens. Figdor's excellent paper on degeneracy in cognitive neuroscience (Figdor 2010) presents empirical research that points in that direction. Considering the uncountable dimensions along which neuroanatomy as well as neurophysiology can be mapped onto cognitive and other supervenient and emergent properties, a heterarchy-inspired approach informed by Simon's comments becomes a reasonable methodological alternative.

The thesis presented in this book is that overlapping constraint regimes exist for a range of organizational levels. The interdependencies that characterize each level are as much the outcomes of context-independent and context-dependent, enabling and governing constraints as they are the effects of forces. Constraints generate indexical phenomena whose invariances must be explained in terms of the contextual constraints shared by the explananda in question. Theories about inflammation, for example, would need to directly address three different but interrelated bodies of law about invariant relations formed by constraints that enable, stabilize, and regulate homeostasis as a whole: one set of homeostatic constraints would be concerned with structure, one with function, and one with functional regulation (Medzhitov 2021). In each pair of levels, the latter supervenes on but exerts governing constraints on the former. Malfunctions of regulatory (governing) constraints can be expected to cascade down and decompensate functional constraints at the next level down. Malfunctions in functional constraints in turn can be expected to make structural constraints go out of kilter, with contextual constraints playing a significant role within and across all levels of organization.

On a positive note, previously unknown constraints shared by vastly different domains might be discovered through such a dimensioned and rich view of multiple realizability. Network theory, for example, has built on those insights to discover constraint architectures shared by widely diverse domains, from terrorist networks to disease transmission patterns.

The interpretive frame of coherence-making by contextual constraints partitions the world along an entirely different set of joints. Discovering shared constraints might reveal a whole new world of interdependencies. It would reveal new dimensions that could explain why two instances from entirely different domains are alike. The following is an illustrative case study.

* * *

The remarkable errors produced by a neural network trained to read words were described earlier. Neural networks and human cases of dyslexia that commit the same type of errors refute Shapiro's claim that the multiple realizability thesis is an a priori argument. These experiments provide empirical evidence of the operation of constraints at the heart of multiple realizability (Juarrero 1999; Moreno and Mossio 2015).

Facial recognition errors committed by both humans and software trained on similar databases offer a related empirical test case for this claim. People commonly misidentify faces of members of the other racial group when it is not represented in the training data set (Bothwell et al. 1989; Castelvecchi 2020; Heyer et al. 2018). If both human cross-racial identification and those of a neural network's middle layers are enabled and governed by shared context-dependent constraints, such errors can be explained as being alike with reference to the type-identified attractors in which both are embedded. Attractors generated by comparable enabling constraints constitute a shared dimension with emergent properties in common that renders two cases ontically alike and not merely coincidental. With respect to that context, each instance is a distinct token of the same Kind of entity. In the facial identification use case, the attractor embodies a multiply realizable type with the emergent property of facial recognition. Realized tokens of this type would be embody those shared constraints that generated the attractor, whether silicon or human.

Constraint regimes define and embody emergent properties of type-level constitutive and governing constraints. On our account, then, constitutive constraint regimes that govern interdependencies of a coherent dynamic describe a shared third dimension with respect to which token outputs are alike; they are tokens of that dynamic type. Attractors formed under the control of a shared constraint regime induced by analogous enabling constraints also illustrate that properties of coordination dynamics and their regulatory constraints carve out reality at different joints from those of folk psychology. Neither our ordinary intuitions nor our classical views of types and kinds would have gathered humans and artificial neural networks under one class. But neither would our ordinary intuitions have classified humans and tornadoes as alike (as dissipative structures).

The invariance in both cases is regulated by a more encompassing set of interdependencies, by the constraint regime of a shared domain that is contextually generated and emphatically not epiphenomenal. As suggested by the attractor concept, self-organized possibility space modifies a system's actions once it is pulled into the attractor's basin. Lesioning human and silicon neural networks below the feedback units—below the constraint regime—reveals the presence of top-down control. It reveals, that is, the organization of complex attractors with effectively comparable constraint architectures in the middle layer of both the neural network and in human association cortices. These interdependencies, generated by the enabling constraints of the training set, especially recursive feedback, can be said to sweep across biological and electromechanical laws and support their alikeness. Critically, however, because they are contextually constrained, the interdependencies hold only under these conditions, in this context, given that training set, and so on.

With contextual specifications spelled out front and center, the approach presented in this book explains why those constraints continue to hold even when different causal mechanisms and material substrates (silicon and wetware) are at work, or when lesioning the network or anatomical malfunction threatens to undermine its unity of type. This perspective offers a more richly dimensioned understanding of realization relations and multiple realizability. Its robustness supports counterfactuals, but only as conditional upon a specified context. Under those conditions, the concept of governing constraint regimes renders top-down causation effective as meaningful, even if only for a range of realizations, under a range of conditions.

This work has proposed that we take seriously the possibility that real relations of constraint generate, constitute, and govern those interdependencies. In addition to looking inward toward components, the logic of explanation appropriate to complex dynamical systems must also look upward (Wimsatt 1974, 1976), to what Lange calls a shared dimension of constraint. Which dimension of constraint, I would add, can bring about effects—even if not as efficient cause. Stated otherwise, explanations of complex systems must look outward to context and backward to history in which those interdependencies were formed and in which they are embedded—as well as inwards to the system's components and local attractors that realize and specify its current possibility space.

Context changes everything.

15

Empirical Research on Delayed Response

Neuroscience Case Studies about Constraints

Empirical observations processed with certain analytic tools support the hypotheses presented in previous chapters. This chapter describes two sets of neuroscience experiments on delayed responses in laboratory animals. The first suggests that after training (context-dependent constraints), brains of macaque monkeys self-organize a coherent neural subspace with task-defined emergent properties. As revealed by a particular statistical technique, this neural subspace structures, initializes, and controls the monkeys' delayed response on each trial run such that it satisfies the task instructions, which vary each time. The second part of this chapter presents a related study that further explains the neural mechanisms that underpin delayed responses. Specifically, this second study concludes that self-organized and task-defined neural subspaces bring about behavior that is appropriate to the task at hand by organizing dynamic attractors in abstract neural state space. A combination of empirical research and simulations uncovers both the formation of such neural attractors and the constraints that control the monkeys' delayed actions and successfully satisfy the task's demands.

* * *

Neural activity preparatory to voluntary behavior has been of interest to neuroscientists attempting to elucidate the contents of the mind's black box. The mind–brain complex, however, did not readily give up its secrets to studies that relied on summing and averaging single-neuron activity. In particular, when experimental conditions require the subjects to integrate and coordinate distinct streams of information, data sets of single-neuron recordings, no matter how large, are so highly heterogeneous (both across neurons and across experimental conditions) that no obvious moment-by-moment correlation with sensory input or motor output can be detected. When carrying out complex but contextually constrained tasks requires "coordination across neurons," single-neuron recordings have proven particularly unsuitable, no matter how large the data sets or fast the processing runs (Cunningham and Yu 2014). When experiments require subjects to internally process a variety of stimuli before performing a task-defined behavior, data sets that sum and average individual neural responses provide no useful insight.

At first, the lack of clear-cut input/output patterns in single-neuron recordings was attributed to noise. Gradually, neuroscientists began to suspect that "this single-neuron complexity may be the realization of a coherent and testable neural mechanism *that exists only [as a dynamical system] at the level of the population*" (Cunningham and Yu 2014, 1503; emphasis added). Testing this hypothesis called for a different approach. As neuroscientists turned to population-level analyses, one class of statistical methods, *dimensionality reduction*, proved particularly useful in uncovering features of brain activity that could not be discerned with classical methods such as averaging responses across trials of single-neuron recordings. Dimensionality reduction, a statistical technique that transforms data from a high-dimensional representation, yielded insights into integrative neurological patterns that subjects organize in response to the constraints of training and the contextual cues of each trial run. Dimensionality reduction can extract those features that matter from noisy raw data.

Dimensionality reduction had previously been used to uncover neural mechanisms of monkeys performing tasks requiring attention. These studies, in contrast, focus on its application to *instructed delay tasks*, those that include an interval between stimulus presentation and the GO cue to initiate behavior. This delay period (after trial run instruction and stimulus presentation but prior to the initiation of motor activity) is called the *preparatory period*. Dimensionality reduction uncovered complex brain activity during that interval that was hidden in the chaos of single-recording data sets.

Mark Churchland's Team

The experimental conditions in both sets of experiments (Churchland, Cunningham, et al. 2012; Kaufman, Churchland et al. 2014; Mante, Sussillo, et al. 2013) are similar. The reaching experiments are described first. Monkeys fitted with electrodes on their motor and premotor cortexes were trained to reach for and touch (or trace) the correct target on a monitor in response to instructions. Some runs required a straight reach, others required a curved reach, some required the monkey to trace out a maze on a screen in a clockwise direction, and on other runs it required a counterclockwise reach. Critically, there was a randomly varying delay interval between the preparatory period and the cue to initiate motion.

Preparatory cortical activity (or motor preparation) was defined as neural activity during the delay period between the instruction and display on the one hand, and the GO cue, on the other. The central question investigated in all these studies is, What exactly does preparatory activity do and how does it segue into actual and task-appropriate responses?

Preparatory and Perimotor Neural Activity

Recordings showed that individual neurons are uncorrelated during the early preparatory period. For approximately the first 200–300 ms, not only does the temporal structure of recordings of individual neurons "var[y] widely across cells: some even increase their firing rate, some decrease, some arrive at an approximate plateau level, while others undulate" (Shenoy, Kaufman, et al. 2011, 38).

It was initially hypothesized that the combined activity of individual neurons during the preparatory period might represent a summing of all the physical dimensions that make up the full data set: distance to the target, direction of the target, or speed of arm motion to the target, the arm's angle of rotation, and so on. However, this turned out not to be so. Few individual neurons appear tuned to these lower-level parameters (Cunningham and Yu 2014; Churchland, Santhanam, et al. 2006; Shenoy, Kaufman, et al. 2011).

Instead, dimensionality reduction revealed population-level neural activity that integrates a wide variety of specific factors to represent the task at hand. That is, during preparatory activity, the monkey constructs an abstract task space in neural space. This abstract space recodes noisy, high-dimensional data into a low-dimensional and complex neural space whose parameters capture salient aspects that covary with relevant elements of the task. The space also initializes (sets the initial conditions) for the subsequent *perimotor stage* of cortical activity, the interval between the GO cue and the actual initiation of movement. The task space organized during the preparatory period is therefore a "dynamical system [that] controls movement" (Churchland, Cunningham, et al. 2012, 51) in the sense that it is predictive of actual reaction times and movement.

Specifically, "Consider the space of all possible preparatory states (all possible Ps). For a given reach, there is presumably some small [contiguous] subregion of space containing those values of P that are adequate to produce a successful [response]" (Shenoy, Kauffman, et al. 2011, 42). That is, it contains those values that would qualify as a successful performance of the required task. Applying dimensionality reduction to massively disparate data sets of individual neural activity during the preparatory period revealed this multidimensional but far smaller subregion of neural space (a task-defined and structured preparatory space) within which subsequent behavior must be constrained to successfully execute the [response] and fulfill the experimental task's demands. As evidenced in the neural activity during the perimotor period, correct responses then flow from this organized preparatory subregion in neural state space into appropriate behavior.

Surprisingly, Churchland's team found that the population-level activity of this organized space consists of rotating and continuous cyclic patterns of neural activity. That is, neural dynamics organized during preparatory activity are rhythmic. The monkey's brain is recoding signals in the frequency domain. Discovering this rotation was surprising because "the [the task-appropriate responses required in the experiments] themselves are not rhythmic" (Churchland, Cunningham, et al. 2012, 51–52).

From the perspective presented in this book, the rhythmic rotations are significant because they remain invariant throughout the trial run: "for each data set the neural state rotates in the same direction across conditions (Churchland, Cunningham, et al. 2012, 53). The authors hypothesize that the continuous properties of the rhythmic activity (such as direction of rotation, amplitude, and phase) capture salient and relevant properties of the task that covary across different experimental conditions. They capture effective input, in our terminology. This conclusion is further supported by the fact that the properties of the rhythmic activity do not correspond to any physical features of the signals displayed. The continuous properties of the constructed task space appear to result "not from a multiphasic signal, but from how that signal is constructed" by the monkey (Churchland, Cunningham, et al. 2012, 55). These population-level patterns of neural activity, I submit, describe intensional content representing the task at hand (reaching for a target, looking at a target, determining if the relevant response is about dot color or dot motion, for example). It implies that the abstract space is eminently multiply realizable and indexically ordered. Indeed, the authors conclude that "many features of the observed rotations make sense in terms of how the actual behavior (EMG, kinematics) *might* be generated in the remainder of the trial run, rather than in terms of the behaviors themselves" (Churchland, Cunningham, et al. 2012, 55; emphasis added).

Rotational invariance of population-level neural activity suggests, in short, that in response to training, neural activity during the preparatory period creates a persistent constraint regime that represents task-defined requirements and anticipates a range of possible movements from which a response will be selected depending on the details of the particular run. It represents the meaningful features of a contextually constrained and type-defined task. The end state of the preparatory period then becomes the initial condition of perimotor activity for that trial run. "If this initial condition is known, subsequent states [the actual behavior on each trial run] can be predicted" (Churchland, Cunningham, et al. 2012, 53). That is, once the GO signal is given, the characteristics of the preparatory abstract space transition naturally through the perimotor neural activity into actual and task-relevant behavior, be it a successful reach or, in a different experiment, a saccade (Shenoy, Kaufman, et al. 2011, 39). It is tempting to conclude from this experiment that, as Patten and Auble proposed, the monkey's cortex generates an abstract model with emergent properties that subsequently constrains actual muscle activity top down such that "individual unit responses . . . reflect the underlying dynamical factors: the patterns present on each axis of [task-defined] space" (Churchland, Cunningham et al. 2012, 54).

A Further Study

In a second set of studies of context dependence in delayed response, rhesus macaque monkeys were presented with a display of noisy signals—in this case, differently colored light dots moving toward the left or right of a monitor. The monkeys were trained to discriminate between the light dots' prevalent direction of motion (called the motion context) and their prevalent color (color context) in light of a contextual cue (task instruction) given prior to the beginning of each trial run. Cued to a color context, for example, the monkeys were trained to select the prevalent color, red or green, of the dots displayed; instructed to a motion context, the monkey had to select their prevalent direction of motion, toward the left or right. The monkeys reported their choices by saccade eye motions to one of two targets placed on either side of the monitor.

At the start of each trial, the monkeys were given instructions about the context for the upcoming run: whether the run would be about color or about motion. Once the contextual cue was provided, the display showed the target options on either side of the screen: red or green for a color context, right or left for motion. The placement of these targets varied randomly on each trial run (sometimes the correct prevalent color target was on the left; at other times, it was on the right).

The monkeys were then presented with a random but noisy display of light dots that were both moving and variously colored. After a few seconds' delay (which, again, varied randomly), the GO cue was given.

The experiment demonstrated that the monkeys were able to discriminate evidence relevant to the instructed context for the task at hand while largely ignoring irrelevant evidence. If the trial run's cued context was about color, the monkeys largely ignored the dots' direction of motion, regardless of their speed and uniformity. If the trial run was about motion, the monkeys largely ignored color, regardless of the display's brightness or color uniformity. Strength of sensory input (how bright the color, for example) did not override relevance, indicating that the behavior was constrained top down, according to the relevant, task-defined contextual cue.

While monkeys performed this task, population-level responses from neurons in a region of the prefrontal cortex (PFC) that selects and executes eye movements, commands visuospatial attention, and integrates information "toward visuo-motor decisions" were recorded. At the start of each run, population-level neural activity converged to a starting baseline located near the center of the plots (as if to be positioned to respond correctly in either context). In the interval after the contextual cue (choose color!) and the target locations (red on right of screen, green on left) were presented but before the dot display was turned on, the monkeys' neural activity quickly moved away from the earlier center baseline and lined up along a neuronal pattern the scientists labeled an *axis of choice* (Mante, Sussillo, et al. 2013, 79).

As in the various experiments conducted by Churchland's team, Mante's so-called axis of choice is a structured, low-dimensional subregion of the monkeys' neural space. Specifically, Mante's team established that this subregion is organized as a line attractor with the options for each choice context located on either end of the axis (green and red in runs with a color context, right and left in runs with a direction of motion context).

Mante, Sussillo, et al. hypothesized that the gradual repositioning of neural activity along an organized axis of choice is evidence that the preparatory period integrates sensory input and contextual cues into a single domain that captures the particular trial run's context; the monkeys' brain activity organized the variables of the task into a coherent neural subregion that defines the requirements of the particular trial run and task-defined response options; the axis of choice spans the task variables, without the values.

The study confirmed this hypothesis: comparing population responses in the PFC across contexts revealed that a single, stable neural pattern integrates sensory evidence in terms of the choices for each particular trial run. The integration combines in one domain the properties relevant to that trial run (dot color, red or green, or prevalent motion toward the right side of the monitor or left).

Choice thus begins by structuring a contextually generated type-defined subregion of possibility space in the PFC: "This will be a color run," or "this will be a direction of motion run," for example. This low-dimensional subspace is organized as the line attractor labeled the axis of choice; it represents the task requirements and the variables for the task: {color [red, green]} or {direction of motion [rightward, leftward]}. This possibility space aligns the options from which to subsequently select.

Significantly for our purposes, the authors of the study note that the results imply that integration and selection are two aspects of a common process separable only at the population level. The study by Mante, Sussillo, et al. thus agrees with experiments conducted by Churchland's team: it proves the construction of an organized neural space that is contextually generated, task-defined, and multiply realizable, it spans possible behaviors that would satisfy that task. This neural space also defines the conditions of satisfaction that will constrain actual behavior top down such that the response is task-appropriate. In Lange's terminology, the axis of choice describes a self-organized third dimension, a regime of constraints that structures a task-defined field, along with the possible range of values for that field. In the language of philosophical action theory, the attractor of the axis of choice might be considered the neural correlate of a *prior intention* to carry out the task of this trial run as instructed by the contextual cue.

Mante, Sussillo, et al. also and simultaneously recorded direct sensory responses to the motion and color displays. At the population level, PFC activity also showed that neural responses to sensory information alone oriented along two separate sensory attractors, one they called an axis of color, the other an axis of motion. Each of these *sensory axes* represented the order parameters of that run: they capture the significant properties (salience, cohesion) of the signals relevant to the contextual cue for that run. That is, they capture information about how strongly coherent the dot display was with respect to the context (color or motion), and how relevant to the contextual cue the features of the dot display were during a given trial run. Neural activity along the axis of motion, for example, was less strong during trial runs cued to a color context—no matter how coherent the direction of motion or how bright the colors.

Interestingly, neural activity along these two sensory axes (motion and color) was momentary; neural activity returned to baseline when the display was turned off. Mante, Sussillo, et al. interpreted such transient sensory neural activity as providing only "momentary evidence . . . in favor of the two choices" (Mante, Sussillo, et al. 2013, 79), color or motion. In contrast, however, neural activity tuned along the axis of choice continued throughout the experiment until the actual response (saccade or reach) was completed. Unlike the sensory axes, the influence of the axis of choice appears to ensure that responses remain true to task by providing top-down governing constraints for the correct response to that trial run. The persistence of the axis of choice in constraining responses despite the intervening delay period can be interpreted as the standing cause of the response.

* * *

Once the task at hand is identified and the corresponding subregion of neural state space is structured accordingly, how did the monkeys select one target rather than another? That is, how does neural activity (now constrained in a lower-dimensioned space that is task-defined) select the correct response for that task on that run (red and not green) and then perform the correct action—looking toward the red target on the right rather than the green one on the left, say? In a simulation, Mante's team found that a different vector participates in this final stage of the process.

When population responses across both axes are combined, the momentary information (red or green, right or left) registered along the sensory axes (for motion and color) and supporting a particular response for a given trial run arcs the neural population activity toward the correct answer—that is, toward the relevant sensory axis, motion or color. That is, the orientation of the combined neural activity that represents the integration of information from both the axis of choice and the relevant sensory axis projects strongly in the direction of the relevant target on the screen and strongly orthogonal to the irrelevant one. It represents the phase transition from many separate streams of information into one coherent set of interdependencies.

Mante, Sussillo, et al. call this coherent and arced aligning of neural activity a *selection vector*. By integrating the run's instructed contextual alternatives with the momentary sensory display of that trial run (as given by the relevant sensory axis), the selection vector encodes relevance, a highly context-dependent parameter. Unlike the orientation of the axis of choice, which is unaffected by the momentary recordings of the sensory stimuli along the sensory axes, the selection and performance of the actual response is therefore controlled by integrating both. Once the monkey must act, integrating all contexts simultaneously (the particular trial run context's axis of choice plus the current and momentary sensory information of the display) constructs and deforms a selection vector in neural state space—and orients it toward the relevant target and away from the irrelevant one. From our perspective, projecting strongly parallel to the relevant signals but orthogonal to the irrelevant ones is evidence that the selection vector's constraint regime integrates information on the display into those invariances previously established by the axis of choice (the constraints that govern the task at hand). In effect, a combination of context-dependent constraints, organized as a selection vector, recodes momentary sensory information in terms of its relevance (for this run, given the task context).

Mante, Sussillo, et al. describe the influence of the selection vector as *context-dependent relaxation*. "This mechanism explains how the same sensory input can result in movement along the line attractor in one context but not the other" (Mante, Sussillo, et al. 2013, 82). The process of context-dependent integration that induces the selection vector explains why a particular saccade is performed—toward the relevant target (displayed on one side of the monitor) and away from the other. And so, the correct (appropriate to this trial run) action is performed.

In contrast with the axis of choice, which determines effective input, the selection vector produces and regulates effective output. Applying dimensionality reduction to population-level responses thus confirmed the construction and presence of what we have called here effective input and output in the monkeys' motor cortex: population-level responses revealed that "identical sensory stimuli . . . can lead to very different behavioral responses depending on context" (Mante, Sussillo, et al. 2013, 78). This is possible only because of the previous Many-to-One self-organization of a type-defined task space.

By conditioning selection and behavioral activation on instructional context, the selection vector might also be considered as the neural correlate of a *proximate intention of action*. Proximate intention integrates the *intensional* content of the intention with motivation to the appropriate behavior such that the intensional content is satisfied.

From the perspective of this book, Mante's team shows how effective input and output are generated and acted on. First, and in response to the enabling constraints of training, neural activity induces a Many-to-One transformation by integrating sensory signals into a coherent set of constrained interdependencies. These organize a type-defined abstract space that encodes continuous population-level properties. Dimensionality reduction reveals population-level neural activity that structures an indexically defined and abstract subregion of neural space. This neural space represents the interdependent type-defined task requirements as perceived by the subject. The neural space is multiply realizable; it spans all possible behaviors that might be required to satisfy those constraints. Once integrated into a coherent dynamic and in view of the constraints it embodies, a specific behavior is then selected and acted on in response to more timely and local constraints (provided by the display pattern of each run and as determined by the selection vector's orientation).

The studies by Mante et al. and Churchland et al. are congruent: the abstract subregion of neural state space organized during the perimotor period of Churchland et al.'s experiments corresponds to the line attractor of Mante, Sussillo, et al.'s axis of choice. In both cases, these are type-defined subregions that encode multiply realizable task-specified governing constraints that control for correct behavior while, top down, allowing it to vary depending on context.

From the perspective presented in this book, furthermore, combining the information represented by the axis of choice with sensory information into a selection vector interlaces semantics and syntax into a coherent set of interdependencies enabled and governed by the real-world constraints of the training and trial run instruction. As embodied in population-level neural patterns, the selection vector issues in meaningful behavior constrained top down by indexically defined interdependencies. The suite of delayed response experiments illustrates how enabling constraints (of training) generate coherent, multiply realizable neural dynamics with emergent (task-defined) properties that evolve into and control task-appropriate actions.

16
Concluding Remarks

So where has this trail led us? The clearing we have opened—and it is only a clearing, not the final destination—is one where interdependencies woven together through multiple constrained interactions turn out to be as fundamental as the allegedly standalone things and indivisible particles that for millennia served as the ontic basis of Western philosophy and modern science. To fully understand the emergence of those interdependencies, we looked to the notion of interlocking constraints to expand our ideas of cause and effect. Interlocking constraints reinstated mereological interactions and relations into our conceptual framework. Doing so also restored the possibility of true coherence and top-down causation, from minds to bodies, from ecosystems to individual species.

The new world revealed through the lens of complexity is congruent with contemporary science. The one prerequisite: conditions must be open and away from thermal equilibrium. Context-independent constraints take conditions far from homogeneity and equilibrium and establish background inhomogeneities. Doing so satisfies this first requirement. Next, context-dependent constraints take conditions away from independence by making previously separate entities conditional upon one another. Working together in open conditions far from equilibrium, context-independent and context-sensitive constraints bring about mutually dependent interactions and relations. The processes now covary because, acting as enabling conditions and constitutive/governing regimes, constraints have woven individual processes and particles into integral wholes. The book calls them interactional types. Parts and wholes, tokens and types, can finally be understood as two aspects of a shared and interdependent dynamic.

A new coherence.

Coherence is not other than individual particles and processes that comprise it; types are not other than the tokens that realize it. Components do not fuse into blocklike things or blend together in an amorphous solvent; but neither are interdependencies unstructured tangled banks. Coherent dynamics that are coordinated across multiple scales and dimensions are not dominance hierarchies; they are distributed heterarchies and holarchies. They manifest as differentiated and heterogenous configurations of matter, energy and information held together at the same scale as well as across scales by principles of mutual constraint satisfaction. Significantly, at each scale the interactants retain their integrity even as they contribute to the generation of an emergent whole. Each aspect of a differentiated coherent dynamic embodies and enacts a distinct set of constrained interdependencies with its own logic of emergent properties and powers.

Coherent parts-to-whole and whole-to-parts relations induced by constraints persist in the face of change. Their patterns are metastable, realized as vectors of complex attractors; they define ecosystems, cultures, mindsets, values, and institutions. They are enacted in social practices and activities and are embodied in technologies and traditions. They span the planet's geology, its atmosphere, its biosphere, and its human worlds.

As understood here, then, coherence generally is grounded in interdependent relata conditioned by the past and the spatiotemporal context in which it is embedded—which interdependence, paradoxically, constrained interactions among its constituents themselves induced. By continuously weaving together streams of energy, mass, and information into more comprehensive spatiotemporal dynamics, contextually constrained relations remember; because they are path dependent, they remember. They also specify, modulate, and regulate actual tokens. Records, registers, and memories become incorporated into yet more expansive interdependencies that hold together in time as well as space. As complexity increases each turn of the spiral serves as a foundation, building block, memory, and precursor of the next production of wondrous new properties and powers.

Darwin's survival of the fittest is as much a process of symbiosis making as it is a zero-sum game.

Newly constituted coordination dynamics are effective; they are neither epiphenomenal nor puppets of their constraints. Among the most important capabilities of coherence-making by constraint is top-down influence, from extended coherent invariances down to individual behavior. Once those coherent invariances involve the complex social organizations of human beings, principles, laws, rituals, myths, and traditions have effects just as forceful actions do, both at the individual and the systemic level—just not as forces. So do their emergent properties: thoughts, beliefs, intentions, values, and feelings. All exercise causality top down, from regimes of constraint that define emergent properties to their manifestation in actions, regulated as cascades of negative feedback and analog decision-making and control. In consequence, the resulting behaviors and actions satisfy the truth conditions set by the emergent properties of those constraint regimes, including those pertaining to cognition and values.

Such real and meaningful top-down causality is possible at all only because ongoing context and path dependence generates flexible, multiply realizable types that span diverse tokens. Emergent properties of contextually generated interdependencies are not transitive across scales or levels of organization; they do not commute. Interactional types as presented in this book are regimes of constraint that must be defined indexically; some must be defined intensionally. The emergent perspective from inside a contextually generated constraint regime is different from the view from outside. Revealed by the filter of constraint, real and effective order parameters can embody the emergent properties of perspective, point of view, and subjectivity. New codes and metrics (including valence, affect, and social and moral values) arise alongside new coherent patterns. Genetics, affect, symbolic language, and ethical norms are codes and metrics that transduce external information into the new interdependencies on the inside. Practices and activities that realize that newly encoded information embody the emergent properties of those interdependencies.

* * *

From their earliest formation, mutually constrained and constraining worlds thus spiral through innumerable iterations of a common theme-and-variation motif. Constrained interactions generate increasingly complex forms. Radically new properties and powers emerge as new interdependencies form. Those emergent properties and powers were not merely waiting to be revealed; evolution is not the unfurling of preestablished potentialities. Each new coherence is a genuine creation, the product of path-dependent multiple constraint satisfaction under open nonequilibrium conditions. Coherence making thus brings about major transformations to emergent dynamics. Constraints make creativity possible; emergent coherent dynamics are the creative outcome of mutually constraining and constrained interactions. The price of such creativity and uniqueness is indetermination and uncertainty. I submit it is a price well worth paying.

Earlier chapters concluded that validated interdependencies persist and evolve because of their contribution to metastability. Epistemically and politically, keeping the conversation going (Rorty1982) is a validated enabling constraint that contributes to metastability. Both abiotic metastability and keeping the conversation going presuppose open conditions such that continuous exchanges of matter, energy, and information with context are facilitated and preserved by enabling and constitutive constraints. In particular, validated coherence presupposes the presence of those context-dependent constraints that enable the emergence of novel and flexible interdependencies while at the same time stabilizing and preserving metastabiity.

Only constraint regimes that are self-determining and self-repairing are fluid, and can continuously update contextual information, promote adaptation and evolution, and remain metastable. These considerations imply that governing constraints of metastable dynamics must be distributed, multiply realizable, flexible, and resilient. They must actively encourage and preserve the diversity and multiple realizability of their components. Individuals must not be dissolved into a homogeneous solution; but they lose some of their complex and novel properties if they ignore the context from which they were emerged. They especially lose their integrity if they are unresponsive to the context in which they are currently embedded.

Only metastable constraint regimes yield validated coherence. This is as true for human beings and their communities as it is for Bénard cells and BZ reactions.

* * *

What can we conclude from all this? Tearing down happens naturally. Creating coherence is an energetically costly process: it takes a burst of entropy for context-dependent constraints to irreversibly produce emergent coordination dynamics. Innumerable identical tokens of the one type is a bogus form of diversity. It tends towards stasis and devolution. Uniformity in social contexts does not keep the conversation going. It takes ongoing background, enabling, and governing constraints to generate and maintain multiply realizable metastable interdependencies in nonequilibrium. Such constraints can enable and facilitate evolution and progress. This recursive "form of causality [previously] unknown to us"[1] informs creativity and individuality, from atoms to humans and their organizations, from dust devils to international human organizations; their coherence emerges from the interdependencies the mereological constraints themselves helped cocreate.

Neither puppets nor absolute sovereigns, human beings and the material and social forms of life they induce are true co-creators of their natural and social worlds. We serve as stewards of the metastability, coherence, and evolvability of both of these worlds. Matter matters. History matters. Social and economic policy matters. Most critically, however, because top-down causality as constraint makes room for meaning and value-informed activities, our choices and actions matter tremendously. In acting, we reveal the variables and the values that really matter to us, individually and to the culture in which we are embedded. We must pay attention to what we pay attention to; to which options we facilitate and promote and which we impede and discard. We must pay particular attention to what we do.

The influence of constraints has been dismissed because they do not bring about change energetically. Because they can be tacit and entrenched, their Escher-like characteristics also make them difficult to track. As background constants that go without saying, they have also been taken for granted. Foregrounding these enabling and governing conditions, so different from but as effective as forceful impacts, has been a central goal of this work.

Facilitating the emergence and persistence of validated coherence, of adaptable and evolvable interdependencies that can continue to form and persist in nature, among human beings and between nature and humankind, is among our most compelling responsibilities. Facilitating the emergence and preservation of a thoroughgoing resilience that affords to both the natural and the human worlds the conditions not only to persist but especially to evolve and thrive is the most pressing moral imperative facing humankind today.

Notes

Chapter 1

1. The paragraph conflates two issues: type identity (What makes me a particular type of entity, a human being?) and individuation (What makes me uniquely me?). These will be disambiguated later.

2. In *Smellosophy*, which is all about the context-dependent and relational character of fragrance, Barwich makes the important distinction between *subjective* and *relational*. The latter is objective.

3. Stanley Salthe at times uses the term *kind* in the sense of specification, a subset, as in "Great Danes are a kind of dog."

4. Ancient philosophers worried about such things. The possibility of reassembling the dismembered limbs of saints (scattered in reliquaries throughout Christendom) would not pose a problem at the Second Coming. Gladiatorial contests, on the other hand, did pose a conceptual puzzle: once digested, unlucky gladiators become an integral component of the lion's in-formed matter.

5. The fact that ecosystems are webs of trophic interdependencies and not internal primary properties undermines even this claim. Outside of their ecological niche, lions in the wild do not survive.

6. *Idealism* proposed that Mind was the more fundamental. *Materialism* claimed that only physical stuff was real.

7. Thinking of Formal cause in terms of a "Principle of Being" is open to the same objection.

8. For that reason, the term *evolution*, which means the unfurling of preexisting potential, was used. Today, we distinguish between development and true evolution, the emergence of new defining traits.

9. Aristotle, *Physics* 194 b17–20 in *The Complete Works*. See also *Posterior Analytics* 71 b9–11:94-a20 in *The Complete Works*.

10. Thanks to Trent Hone for the following observation: "hammer" is a multiply realizable type. Without modification, the term is usually interpreted to mean claw hammer, one realization of the type.

11. See Juarrero and Rubino (2008) for an overview of Emergentism prior to 1950. For an overview of contemporary Emergentism post 1950, see Bedau and Humphreys (2008).

12. In the twentieth century, this issue was readdressed in terms of *projectible predicates* and support of counterfactuals (see chapter 14).

Chapter 2

1. Time and history are included in context.

2. Such phase transitions are transitions to new forms of order. Order parameters "measure the degree of order across the boundaries."

3. *Order parameters* describe systemwide interdependencies of complex systems. For example, *degree of cohesion* of a tornado or a culture is an order parameter that describes a systemwide property of its relations and interactions.

4. Existentialism's emphasis on innate freedom must be reframed as autonomous (self-governed) interdependencies brought about by self-constraint and self-determination. Freedom is not independence from external influences; it is self-rule in community.

Chapter 3

1. To this day, scientists have not discovered how mistletoe (whose genome is twenty-four times the size of the human genome) taps the energy it needs to survive. This is particularly puzzling since mistletoe lacks the mitochondria living things rely on to produce adenosine triphospate (ATP).

2. As suggested earlier in passing, context-dependent constraints, working against a background of context-independent constraints such as gradients, are responsible for generating complex flow *designs* and *patterns* (Bejan 2020). These novel patterns tap otherwise inaccessible energy gradients. This is as true for galaxy and star formation, lightning discharges, and whirlpool eddies as for the organization of human societies. All order creation satisfies the second law; different varieties of order emerge with different varieties of constraints.

Chapter 4

1. Juarrero (1999) described these constraints as *context free*. That characterization was misleading; it elicited criticisms to the effect that nothing is entirely context free except the universe. This work instead uses *context independent* to mean what was previously meant by context free—that is, distance from equilibrium. Context-independent constraints set up possibility landscapes.

2. "The quality or condition inherent in a body that exhibits opposite properties or powers in opposite parts or directions or that exhibits contrasted properties or powers in contrasted parts or directions" (Merriam-Webster).

3. Probability distribution is the set of relative likelihoods that a variable will have a value in a given interval.

4. I submit that radical *sensemaking* (Snowden 2015) calls for reimagining novel context-independent constraints that can reshape a qualitatively novel and coherent possibility space.

5. You can only "play" fifty-two-card pickup with indistinguishable playing cards. "Fifty-two-card pickup" is therefore a joke, not a game.

6. The past history hypothesis of the origin of the universe addresses this issue: if the universe must have originated in conditions of low entropy (a potential energy hill), what is the minimum height it could have been? How low must/can that entropy have been?

7. For discussions about multidimensional epigenetic landscapes, see Jablonka and Lamb 2002.

8. They do not strictly determine the future because we are in a multidimensional landscape with multiple constraint satisfaction and top-down governing constraints that constrain in virtue of their emergent properties and not mere energetic considerations.

9. Vague concepts such as baldness and turquoise (the shade of color) have fuzzy boundaries, where they stop applying is unclear. Ambiguous concepts have two or more different meanings: "blue" can mean a color or a mood.

Chapter 5

1. The rest of this chapter draws heavily from Halloran and Struchiner (1991).

Chapter 6

1. See https://www.pbs.org/lifebeyondearth/resources/intkauffmanpop.html.

2. Adding degrees of freedom is characteristic of these dynamics.

3. In all cases, we refer to open systems that exchange matter and energy with their environment. However, not all are alike: the initial and boundary conditions of convection cells are externally set by the scientist conducting an experiment or, in the case of hurricanes, by the atmospheric conditions. This changes with the emergence of autocatalytic cycles, whose endogenous dynamics generate and regenerate the very conditions they require to form and persist.

4. The transition also marks the appearance of a hierarchical form of organization.

5. As always, context includes time and history as well as natural and psychosociocultural conditions.

6. As a curious aside, the adoption of one constraint (a social practice) impedes, top-down, the activation of another constraint (an enzyme).

7. This may turn out to be premature. Layered biofilms might show incipient organization, suggesting that *stacking* might be an early context-dependent constraint.

8. Pascal and Pross (see discussion in chapter 9) dismiss the need to stabilize exponential growth. In our view, enabling constraints induce mereological dynamics, which are then stabilized by constitutive and governing constraints. Thanks to the nonlinearities involved, they keep actual tokens true to type through cascades of negative feedback from the overarching constraint regime to the token realizations.

9. This distinction might have been what Jean Jacques Rousseau was seeking in his discussion of a general will.

Chapter 7

1. Enzymes are catalysts that facilitate individual biochemical reactions.

2. The latter requirement means the water cycle is not autopoietic.

3. According to German philosopher Immanuel Kant, this recursive "form of causality unknown to us" is the foundation of end-directed behavior (Juarrero-Roque 1985).

4. I will ignore their claim that agency originates with life (see Salthe 2015 review of *Biological Autonomy*).

5. Contrary to Shannon's interpretation of information, Pattee (1982) considers *semantic* and *informational* synonymous.

Chapter 8

1. Mechanism is used here, not in the sense of a machine part but in the sense of a process, or a way of acting and behaving that helps achieve a particular result.

2. Exemplifying a precursor of teleology, mitosis is in the service of the Type for as long as governing constraints hold. Runaway cell division that characterizes cancer escapes that control and threatens the organism's survival.

3. I will not examine inherited constraints due to epigenetic influences or recombinant DNA; nor will I examine metagenomic constraints of the microbiome. My conjecture is that over time, these reconfigure effective input to the organism.

4. The only source of variation would be epigenetic influences on an organism's path-dependent trajectory.

5. With few exceptions such as reports of atypical behavior of caged organisms under crowded conditions, density and packing are rarely studied for either their enabling or restrictive capabilities. This even though, by increasing the likelihood of the appearance of context-dependent constraints, cramped conditions can facilitate *hybridization* (Guo et al. 1991), a source of new species.

6. Context dependence, including path dependence, makes exact prediction impossible; it is the ultimate source of Kauffman's unprestatability.

7. Authors of utopias understood the need to isolate perfect societies to increase their likelihood of survival (Juarrero 1991). Utopian literature is not part of South or East Asian cultures; Buddhism and Taoism are centered around flow and context, not unchanging and internal universal essences.

8. Seed dispersion by birds and pollen dispersion by insects are early mechanisms that extend the range of a plant's constraint regime; they are precursors to organismic motility. Research on mychorrizal networks (networks of matted fungal filaments that wrap around and penetrate tree roots) suggests the presence of symbiotic interdependencies between different kingdoms.

9. See https://cynefin.io/images/e/e5/CognitiveEdge_Scaffolding.pdf.

10. Searle's intention in action would apply to the prosthesis as it did to the original organic limb.

Chapter 9

1. In isolated systems like the universe, inhomogeneity in the initial conditions is unaccounted for (Barbour 2020). In any case, left to themselves, only the final scene, once the play has run its course, lasts.

2. Or explained away by a nominalist stance.

3. Ice ages and desertification occur when those governing constraints weaken.

4. As an origin story, stumbling upon a stable configuration is no different from evolution's concept of random mutation. In our view, a combination of context-independent and context-dependent constraints is a systematic source of metastable configurations.

5. As we have been arguing, this possibility presupposes the operation of constraints. But as noted, nontrivial and long-lasting distinctions between macro- and microstates suggest the presence of governing constraints, not mere statistical improbability.

6. The terms are ambiguous; by *robustness*, we mean a system's ability to resist or withstand change without having to adapt. *Resilience*, in contrast, is used to mean a system's ability to survive by adapting.

Chapter 11

1. Allostasis describes the processes that adjust and harmonize individual dynamics such as heart and blood pressure and maintain homeostasis, for example.

2. "Overlapping habitats" are emergent niches that result from "constructive interference" between two previously independent habitats.

3. In recent decades, for example, network theory has shown that network and topological properties of terrorist networks and electrical grids alike can share similar constraint architectures (Barabasi, Newman, and Watts 2006).

4. "Parkour squirrels" seem to learn to adjust their position in midair in response to interoceptive and proprioceptive signals (Hunt et al. 2021).

5. Although social organization emerges from the workings of contextually constrained interactions among individuals, once the society's constraint regime coalesces, its order parameters establish a new context within which behavior will subsequently be controlled. Ethical and moral values are therefore context dependent as second-order context-dependent constraints. They establish the context within which actions are performed and regulated.

6. It is important to note that he was referring to biological self-regulating processes like homeostasis.

7. In the prebiotic world, the description "long-term projects" would have equated to top-down analog control in terms of continuous metastable dynamics as described earlier.

8. Consider your dining room chandelier's light switch: a quick binary (digital) toggle turns it on or off. Analog dimmer switches, in contrast, support a continuous range of brightness levels, as desired, under different conditions.

9. This is the main stumbling block of AI, even including machine learning (Mitchell 2021). It does not generalize to other domains and as a result is brittle.

10. Climate Home News, a part of the Guardian Environment Network, reported this in December 2017. https://www.theguardian.com/environment/2017/dec/11/tsunami-of-data-could-consume-fifth-global-electricity-by-2025#:~:text=4%20years%20old-,'Tsunami%20of%20data'%20could%20consume%20one%20fifth,of%20global%20electricity%20by%202025&text=In%20an%20update%20to%20a,world's%20carbon%20emissions%20by%202025.

11. Recent research on "infant metaphysics" (Thomas et al. 2022) reveals that young children's willingness to share saliva with those with whom they have intimate bonds suggests that they interpret their world in terms of thick relationships ("attachments, obligations, and mutual responsiveness").

12. Earlier in chapter 7, we noted Pattee's analogous point about semantic interpretation.

13. This account is compatible with theories of the mind being extended, embodied, and enacted (Clark 1996; Chemero 2009; and Di Paolo, Buhrmann, and Barandiaran 2017). See chapter 13.

Chapter 12

1. One noteworthy exception is Trent Hone's (2018) account of information management in the U.S. Navy.

2. Hierarchical theorists from the mid-twentieth century use the term *hierarchical control* to refer to top-down control.

3. That understanding of the origin of life is congruent with the views presented by Montevil, Moreno, and Mossio mentioned earlier. Holons are coherent interdependencies, so critiques of the principle of persistence presented in chapter 9 stand.

Chapter 13

1. C-fibers were big at the time. Ironically, C-fibers are afferent nerves in the peripheral sensory system—not in the brain as such. Despite its significance to type identity, that detail was somehow never mentioned.

2. This prompted American psychologist Karl Lashley to espouse *totipotentiality*, the (quickly discredited) capacity of *any* physical brain element (neurons) or process to produce any mental event.

3. Since "brain events" referred to organic wetware, the psychoneural type identity theory was opposed by the AI community on the grounds that it barred even the possibility of silicon-based intelligence.

4. Chalmers holds a double aspect theory today.

5. Carl Gillett (2016) argues that, unlike philosophers, physicists like Steven Weinberg and Bryan Laughlin routinely acknowledge causation by collectives that are not other than their components—and of which Gillett offers the standard

examples such as lasers, homeostasis, superconductivity and coordinated marching. However, Gillett does not directly address the causes of these collectives: what generates and differentiates them from mere clumps and masses, or confers on them their novel collective properties. In particular, he does not address the source of their capacity to preserve the collectivity—the interdependences—that distinguish them from agglomerations over time despite turnover of components.

Chapter 14

1. The role of hormones in the brain's information-processing capacity makes this a highly questionable statement.

2. At what level of granularity (Bechtel and Mundale 1999) is "causally relevant mechanism" defined? At what level of granularity must "causally relevant properties" bring about the same result but "in a different way"? Why single lens and retina in octopi and mammals . . . but not the compound eyes of arthropods?

3. From the *Oxford Dictionary of Philosophy*, https://www.oxfordreference.com/view/10.1093/oi/authority.20110803100348986.

4. Note retreat from causal laws to "principles."

5. Formulating intertheoretic bridge laws from one (reducing) field of base entities to another (reduced) field of functional entities would be necessary (Darden 1977). The relations between the entities in the reducing and reduced theory, we submit, would be relations of constraint.

6. This is precisely the argument of small countries like Barbados, whose prime minister notes that focusing solely on the boundaries and coordinates of the global economy's possibility space fails to take seriously the actual topology of micro contexts within that rugged landscape, where context-dependent constraints like climate change's influence on the severity and frequency of hurricanes wreak havoc on debt obligations.

7. Lange uses Richard Feynman's (1967, 59) term.

8. Lange does mention spatial dimensionality: whereas Kant suggests inverse square laws that explain why space is three-dimensional, Hertz proposed that inverse square laws are explained by three-dimensional space: the latter's constraints limit what laws exist: factual inverse square laws are those that are compatible with the constraints of three-dimensional space.

Chapter 16

1. Here I am referring to a concept explored by Immanuel Kant. For further discussion of this concept, see Juarrero-Roque 1985.

References

Abbott, A. 2018. How the brain's phase code might unlock the mysteries of perception. *Nature*, December 11.

Abraham R., and C. Shaw. 1992. *Dynamics: The geometry of behavior*. New York: Basic Books.

Adam, D. 2020. Walk this way. *NewScientist*, September 10, 36–40.

Agosta, S. J., and D. R. Brooks. 2020. *Major metaphors in evolution: Darwinism then and now*. New York: Springer.

Agosta, S., N. Janz, and D. R. Brooks. 2010. How specialists can be generalists: Resolving the parasite paradox and implications for emerging infectious diseases. *Zoologia* 27: 171–162. https://www.researchgate.net/publication/252628776_How_Specialists_Can_Be_Generalists_Resolving_the_Parasite_Paradox_and_Implications_for_Emerging_Infectious_Disease.

Ahmadian, Y., and K. Miller. 2021. What is the dynamical regime of cerebral cortex? *Neuron* 109:373–391. https//doi.org/10.1016/j.neuron.2021.07.031.

Allen, T. F. H., R. V. O'Neill, and T. W. Hoekstra. 1984. *Interlevel relations in ecological research and management: Some working principles from hierarchy theory*. USDA Forest Service Technical Report RM-110.

Allen, T. F. H., and T. B. Starr. 1982. *Hierarchy*. Chicago: University of Chicago Press.

Anderson, Philip W. 1972. More is different. *Science* 177(4047):393–396. https://doi.org/10.1126/science.177.4047.393.

Aristotle. 2022. The complete works. ATN Classics. ASIN: B08X4W3GVJ.

Arnellos, A., and A. Moreno. Cognitive functions are not reducible to biological ones: the case of minimal visual perception. *Biology and Philosophy* 37:1–25.

Arthur, B. 2020. Why does economics need this different approach? In *Complexity economics: Dialogues of the applied complexity network*, edited by B. Arthur, E. D. Beinhocker, and A. Stanger. Santa Fe, NM: SFI Press.

Arthur, B., E. D. Beinhocker, and A. Stanger, eds. 2020. *Complexity economics: Dialogues of the applied complexity network*. Santa Fe, NM: SFI Press.

Artigiani, R. 1994. Send me your refuse: The US Constitution as trash collector. *American Journal of Semiotics* 11(1/2):249–276.

Artigiani, R. 2021. Shifting paradigms: Beyond modern science to complexity and ethics. *Northern Plains Ethics Journal* 9(1):1–98.

Ashby, W. R. 1958. Requisite variety and its implications for the control of complex systems. *Cybernetica (Namur)* 1(2):1–13.

Aznar, A. 2021. Why psychologists can't decide if moral disgust is even a thing. *NewScientist* 3356. https://www.newscientist.com/article/mg25133564-300-why-psychologists-cant-decide-if-moral-disgust-is-even-a-thing/.

Barabasi, A.-L., M. Newman, and D. J. Watts. 2006. *The structure and dynamics of networks*. Princeton, NJ: Princeton University Press.

Barandiaran, X. 2008. Mental life: A naturalized approach to the autonomy of cognitive agents. PhD thesis, University of the Basque Country, Donostia-San Sebastian, Spain. https://xabierbarandiaran.files.wordpress.com/2010/06/barandiaran-2008-phd_thesis.pdf.

Barber, N. 2016. Why entrench? *International Journal of Constitutional Law* 14(2):325–350. https://doi.org/10.1093/icon/mow030.

Barbour, J. 2020. *The Janus point: A new theory of time*. London: The Bodley Head.

Barwich, A. S. 2020. *Smellosophy: What the nose tells the mind*. Cambridge, MA: Harvard University Press.

Bateson, G. 1960. *Mind and nature: A necessary unity*. New York: Ballantine Books.

Bateson, G. 1972. *Steps to an ecology of mind*. New York: Ballantine.

Bechtel, W., and J. Mundale. 1999. Multiple realizability revisited: Linking cognitive and neural states. *Philosophy of Science* 66:175–207.

Bedau, M. and P. Humphreys, eds. 2008. *Emergence: Contemporary readings in the philosophy of science*. Cambridge, MA: MIT Press.

Bejan, A. 2016. *The physics of life: The evolution of everything*. New York: St. Martin's Press.

Bejan, A. 2020. Discipline in thermodynamics. *Energies* 13:2487.

Bejan, A., and S. Lorente. 2004. The constructal law and the thermodynamics of flow systems with configuration. *International Society of Heat and Mass Transfer* 47(14–16):3203–3214.

Bejan, A., and S. Lorente. 2008. *Design with constructal theory*. New York: Wiley.

Bénard, H. 1900. Les tourbillons cellulaires dans une nappe liquide [Cellular vortices in a sheet of liquid]. *Revue Générale des Sciences Pures et Appliquées* 11:1261–1271, 1309–1328.

Bich, L., M. Mossio, and A. M. Soto. 2020. Glycemia regulation: From feedback loops to organizational closure. *Frontiers in Physiology* 11:69. https://doi.org/10.3389/phys2020.00069.

Bickhard, M. 1992. Scaffolding and self-scaffolding: Central aspects of development. In *Children's development within social contexts*, edited by L. T. Winegar and J. Valsiner, 33–52. Hillsdale, NJ: Lawrence Erlbaum.

Bloom, J. 2015. Cucumbers and fish—what is called culture. https://vimeo.com/145109744.

Bonduriansky, R., and T. Day. 2020. *Extended heredity: A new understanding of inheritance and evolution*. Princeton, NJ: Princeton University Press.

Bothwell, R.K., Brigham, J.C., Malpass, R.S. 1989. Cross-racial identification of faces. *Personality and Social Psychology Bulletin* 15:19–25.

Bourdieu, P. 1977. *Outline of a theory of practice*. Cambridge: Cambridge University Press.

Bourdieu, P. 1986. *Distinction*. New York: Routledge Classics.

Brahic, M. 2021. The Jekyll and Hyde proteins. *NewScientist* 252:48–50.

Brillouin, L. 1962. *Science and information theory*. New York: Academic Press.

Brooks, D. R. 2010. Sagas of the children of time: The importance of phylogenetic teaching in biology. *Evolution: Education and Outreach* 3:495–498.

Brooks, D. R., and E. O. Wiley. 1988. *Entropy as evolution*. 2nd ed. Chicago, IL: University of Chicago Press.

Buchler, J. 1977. Notes on a contour of natural complexes. In *Metaphysics of natural complexes*, edited by J. Buchler. New York: SUNY Press.

Burgauer, M. 2022a. Condensates. Maturity Mapping. https://maturitymapping.com/constraintregimes.

Burgauer, M. 2022b. Practice theory. https://maturitymapping.com.

Campagna, L., and S. Turbek. 2021. The geography of speciation in the Ibera seedeater. *Science*. https://www.science.org/do/10.1126/comment.762207/full/.

Campbell, D. 1974. Downward causation. In *Studies in the philosophy of biology*, edited by F. Ayala and T. Dobzhansky, 179–186. Berkeley: University of California Press.

Caporael, L. R., J. R. Griesemer, and W. C. Wimsatt, eds. 2014. *Developing scaffolds in evolution, culture, and cognition*. Cambridge, MA: MIT Press.

Carello, C., and M. T. Turvey. 2002. The ecological approach to perception. In *Encyclopedia of cognitive science*, edited by L. Nadel. London: Nature Publishing Group.

Castelvecchi, D. 2020. Beating biometric bias. *Nature* 587:347–349.

Centola, D. 2018. *How behavior spreads: The science of complex contagions*. Princeton, NJ: Princeton University Press.

Chalmers, D. 1996. *The conscious mind*. Oxford: Oxford University Press.

Chemero, A. 2003. An outline of a theory of affordances. *Ecological Psychology* 15(2):181–195.

Chemero, A. 2009. *Radical embodied cognitive science*. Cambridge, MA: MIT Press.

Chemero, A., and M. Turvey. 2007. Gibsonian affordances for roboticists. *Adaptive Behavior* 12:473–480.

Churchland, M. M., J. P. Cunningham, M. T. Kaufman, J. D. Foster, P. Nuyujujian, S. I. Ryu, and K.V. Shenoy. 2012. Neural population dynamics during reaching. *Nature* 487(7405):51–56. https://doi.org/10.1038/nature11129.

Churchland, M. M., G. Santhanam, and K. V. Shenoy. 2006. Preparatory activity in premotor and motor cortex reflects the speed of the upcoming reach. *Journal of Neurophysiology* 96:3130–3146.

Churchland, M. M., and K. V. Shenoy. 2007. Temporal complexity and heterogeneity of single-neuron activity in premotor and motor cortex. *Journal of Neurophysiology* 97:4235–4257.

Churchland, M. M., B. M. Yu, S. I. Ryu, G. Santhanam, and K. V. Shenoy. 2006. Neural variability in premotor cortex provides a signature of motor preparation. *The Journal of Neuroscience* 26:3697–3712.

Churchland, M. M., B. M. Yu, M. Sahani, and K. V. Shenoy. 2007. Techniques for extracting single-trial activity patterns from large-scale neural recordings. *Current Opinion in Neurobiology* 17:609–618.

Cilliers, P. 1998. *Complexity and postmodernism: Understanding complex systems*. London: Routledge.

Clark, A. 1996. *Being there: Putting brain and world together again*. Cambridge, MA: MIT Press.

Clark, A. 2019. *Surfing uncertainty: Prediction, action, and the embodied mind*. Oxford: Oxford University Press.

Clark, A., and D. Chalmers. 1998. The extended mind. *Analysis* 58(1):7–19.

Co, A. D., and M. P. Brenner. 2020. Tracing cell trajectories in a biofilm. *Science* 369(3031):6488. https://doi.org/10.1126/science.abd1225.

Collier, J. 2003a. Fundamental properties of self-organization. In *Causality, emergence, self-organization*, edited by V. Arshinov and C. Fuchs, 150–166. Moscow: NIA-Piroda.

Collier, J. 2003b. Hierarchical dynamical information systems with a focus on biology. *Entropy* 5:100–124.

Collier, J. 2004. Self-organization, individuation and identity. *Revue Internationale de Philosophie* 59:151–172.

Collier, J. 2005. Change and identity in complex systems. *Ecology and Society* 10(1):29. http://www.ukzn.ac.za/undphil/collier.

Collier, J. 2010. A dynamical approach to identity and diversity in complex systems. In *Complexity, difference and identity: An ethical perspective*, edited by P. Cilliers. New York: Springer.

Collier, J., and C. Hooker. 1999. Complexly organised dynamical systems. *Open Systems and Information Dynamics* 6(3):241–302. https://doi.org/10.1023/A:1009662321079.

Collier, J., and S. J. Muller. 1998. The dynamical basis of emergence in natural hierarchies. In *Emergence, complexity, hierarchy, organization, selected and edited papers from the Echo III Conference*, edited by G. L. Farre and T. Oksala. Helsinki: Acta Polytechnica Scandinavica.

Conrad, M. 1972. The importance of molecular hierarchy in information processing. In *Towards a theoretical biology*, edited by C. H. Waddington. Edinburgh: Edinburgh University Press.

Conrad, M. 1983. *Adaptability: The significance of variability from molecule to ecosystem.* New York: Plenum Press.

Conway, G. H. 1970. www.playthegameoflife.com. *Scientific American* 223:120–123. https://doi.org/10.1038/scientificamerican1070–120.

Crick, F., and C. Koch. 1990. Toward a neurobiological theory of consciousness. *Seminars in Neuroscience* 2:264–275.

Cumhaill, C., and R. Wiseman. 2022. *Metaphysical animals: How four women brought philosophy back to life.* New York: Doubleday.

Cunningham, J. P., and B. M. Yu. 2014. Dimensionality reduction for large-scale neural recordings. *Nature Neuroscience* 17(11):1500—1509.

Damasio, A. 2021. *Feeling and knowing: Making minds conscious.* New York: Pantheon.

Darden, L. 1977. Interfield theories. *Philosophy of Science* 44:43–64.

Datar, A., and N. Nicosia. 2018. Assessing social contagion in body mass index, overweight, and obesity using a natural experiment. *JAMA Pediatrics* 172(3):239–246.

Davidson, D. 1980. Mental events. In *Essays on actions and events*, 207–225. Reprint, Oxford: Clarendon Press.

Dean, M. 2020. "Religion as physics." https://u.pcloud.link/publink/show?code=XZJ4QLXZtxvmhLoA6QjrUj2jtrw1bz5S4m47, video. UCLA Center for the Study of Religion, 36:40.

Dehaene, S., and L. Naccache. 2000. Towards a cognitive neuroscience of consciousness: Basic evidence and a workspace framework. *Cognition* 79:1–37.

De Jaegher, H., and E. Di Paolo. 2007. Participatory sense-making: An enactive approach to social cognition. *Phenomenology and the Cognitive Sciences* 6(4):485–507. https://doi.org/10.1007/s11097-007-9076-9.

Demming, A. 2020. Analogue comeback. *NewScientist*, July 25, 37–40.

Depew, D., and B. Weber. 1995. *Darwinism evolving: Systems dynamics and the genealogy of natural selection.* Cambridge, MA: MIT Press.

Di Paolo, E., T. Buhrmann, and X. Barandiaran. 2017. *Sensorimotor life: An enactive proposal.* Oxford: Oxford University Press.

Donald, M. 1991. *Origins of the modern mind.* Cambridge, MA: Harvard University Press.

Donne, J. 1611. *An anatomy of the world.* https://www.bartleby.com/357/169.html.

Dreyfus, H. 1972. *What computers can't do: The limits of artificial intelligence.* Cambridge, MA: MIT Press.

Dyson, F. 2001. Is life analog or digital? Edge.org. https://www.edge.org/conversation/freeman_dyson-is-life-analog-or-digital.

Dyson, G. 2011. What scientific concept would improve everybody's cognitive toolkit? Edge.org. https://www.edge.org/response-detail/10105.

Dyson, G. 2012. *Turing's cathedral.* New York: Vintage.

Dyson, G. 2019. edge.org. https://www.edge.org/conversation/george_dyson-child hoods-end.

Earley, J. 1981. Self-organization and agency. In chemistry and process philosophy. *Process Studies* 11:242–258.

Edelman, G. 1987. *Neural Darwinism*. New York: Basic Books.

Edelman, G., and G. Tononi. 2000. *A universe of consciousness: How matter becomes imagination*. New York: Basic Books.

Egbert, M., and X. Barandiaran. 2014. Modeling habits as self-sustaining patterns of sensorimotor behavior. *Frontiers in Human Neuroscience* 8:1–15.

Eigen, M. 1971. Self-organization of matter and the evolution of biological macromolecules. *Die Naturwissenschaften* 58(10):465–523.

Eldredge, N. 2015. *Eternal ephemera*. New York: Columbia University Press.

Ellis, G. F. R. 2012. On the limits of quantum theory: Contextuality and the quantum-classical cut. *Annals of Physics* 327:1890–1932.

Ellis, G. F. R. 2016. *How can physics underlie the mind?* New York: Springer.

Ellis, G. F. R. 2021. Physical, logical, and mental top-down effects. In *Top down causation and emergence*, edited by J. Voosholz and M. Gabriel. Cham, Switzerland: Springer.

Etxeberria, A., and J. Umerez. 2013. Organization. In *Encyclopedia of systems biology*, edited by W. Dubitsky, O. Workenhauer, H. Yokata, and H. Cho. New York: Springer. https://doi.org/10.1007/978-1-4419-9863-7_77.

Feigl, H. 1958. The "mental" and the "physical." In *Concepts, theories and the mind-body problem*, edited by H. Feigl, M. Scriven, and G. Maxwell. Vol. 2. Minneapolis: Minnesota Studies in the Philosophy of Science.

Feynman, R. 1967. *The character of physical law*. Cambridge, MA: MIT Press.

Figdor, C. 2010. Neuroscience and the multiple realization of cognitive functions. *Philosophy of Science* 77(3):419–456.

Fodor, J. 1981. What psychological states are not. In *Representations*, 79–99. Cambridge, MA: MIT Press.

Freeman, W. J. 1991a. Nonlinear dynamics in olfactory information processing. In *Olfaction*, edited by J. Davis and H. Eichenbaum. Cambridge, MA: MIT Press.

Freeman, W. J. 1991b. The physiology of perception. *Scientific American*, February, 78–85.

Friston, K. The free-energy principle: A unified brain theory. *Nature Reviews Neuroscience* 11:127–138.

Friston, K., J. G. Tononi, G. N. Reeke, O. Spons, and G. M. Edelman. 1994. Value-dependent selection in the brain: Simulation in a synthetic neural model. *Neuroscience* 59(2):229–243.

Gardner, M. 1970. The fantastic combinations of John Conway's new solitaire game "life" (PDF). Mathematical Games. *Scientific American* 223(4):120–123. https://doi.org/10.1038/scientificamerican1070-120.

Gare, A. 2019. Biosemiosis and causation. Defending biosemiotics through Rosen's theoretical biology, or integrating biosemiotics and anticipatory systems theory. *Cosmos and History* 15(1):31–90.

Gatlin, L. 1972. *Information and the living system.* New York: Columbia University Press.

Gershenson, C. 2020. Guiding the self-organization of cyber-physical systems. *Frontiers in Robotics and AI*, April 3, 2020. https://doi.org/10.3389/frobt.2020.00041.

Gibson, J. J. 1975. Affordances and behavior. In *Reasons for realism: Selected essays of James J. Gibson*, edited by E. S. Reed and R. Jones, 410–411. Hillsdale, NJ: Lawrence Erlbaum.

Gilbert, S. 1991. Epigenetic landscaping: Waddington's use of cell fate bifurcation diagrams. *Biology and Philosophy* 6:135–154.

Gill, M., and J. Lennox, eds. 1994. *Self-motion: from Aristotle to Newton.* Princeton, NJ: Princeton University Press.

Gillett, C. 2003. The metaphysics of realization, multiple realizability, and the special sciences. *Journal of Philosophy* 100(11):591–603.

Gillett, C. 2016. *Reduction and emergence in science and philosophy.* Cambridge: Cambridge University Press.

Goldman, A. 1970. *A theory of human action.* Englewood Cliffs: Prentice Hall.

Goodwin, B. C., and M. H. Cohen. 1969. A phase shift model for the spatial and temporal organization of developing systems. *Journal of Theoretical Biology* 25(1):49–107.

Gould, S.J., and R.C. Lewontin. 1979. The spandrels of San Marco and the Panglossian paradigm: A critique of the adaptationist programme. *Proceedings of the Royal Society of London B* 205:581–598.

Grene, M. 1974. *The understanding of nature: Essays in the philosophy of biology.* Dordrecht, Netherlands: Springer.

Grobstein, C. 1973. Hierarchical order and neogenesis. In *Hierarchy theory: The challenges of complex systems*, edited by H. H. Pattee. New York: Braziller.

Guo, P. Z., L. D. Mueller, and F. J. Ayala. 1991. Evolution of behavior by density-dependent natural selection. *Proceedings of the National Academy of Science* 88(23):10905–10906. https://doi.org/10.1073/pnas.88.23.10905.

Haken, H. 1996. Slaving principle revisited. *Physica D: Nonlinear Phenomena* 97:95–103.

Haken, H., and J. Portugali. 2021. *Synergetic cities: Information, steady state and phase transition.* New York: Springer.

Haken, H., and A. Wunderlin. 1988. The slaving principle of synergetics—an outline. In *Order and chaos in nonlinear physical systems, edited by S. Lundqvist, N.H. March, and M.P. Tosi*, 457–463. Boston: Springer.

Halloran, M. E., and C. J. Struchiner. 1991. Study designs for dependent happenings. *Epidemiology* 2(5):331–338.

Halpern, P. 2020. *Synchronicity: The epic quest to understand the quantum nature of cause and effect*. New York: Basic Books.

Hatna, E., and I. Benenson. 2012. The Schelling model of ethnic residential dynamics: Beyond the integrated-segregated dichotomy of patterns. *Journal of Artificial Societies and Social Simulation* 15(1):6. https://jasss.soc.surrey.ac.uk/15/1/6.html.

Henrich, J. 2016. *The secret of our success: How culture is driving human evolution, domesticating our species, and making us smarter*. Princeton, NJ: Princeton University Press.

Heyer, R., C. Semmier, and A. Hendrickson. 2018. Humans and algorithms for facial recognition: The effects of candidate list length and experience on performance. *Journal of Applied Research in Memory and Cognition* 7:597–609.

Hinton, G. E., and T. Shallice. 1991. Lesioning an attractor network: Investigations of acquired dyslexia. *Psychological Review* 98(1):74–95.

Hinton, G. E., and R. R. Salakhutdinov. 2006. Reducing the dimensionality of data with neural networks. *Science* 313(5786):504–507.

Hoffmann, P. 2012. *Life's ratchet: How molecular machines extract order from chaos*. New York: Basic Books.

Hoffmeyer, J.-H. 2017. Basic biological anticipation. In *Handbook of anticipation*, edited by R. Poli. Cham, Switzerland: Springer International.

Hoffmeyer, J.-H. 2018. Causation, constructors, and codes. *Biosystems* 164: 121–127.

Hofstadter, D. 1979. *Gödel, Escher and Bach*. New York: Basic Books.

Holmes, B. 2019. The Goldilocks planet. *NewScientist* 24(3222):34–37.

Hone, T. 2018. *Learning war: The evolution of fighting doctrine in the U.S. Navy, 1898–1945*. Annapolis, MD: Naval Institute Press.

Humphries, M. 2021. *The spike: An epic journey through the brain in 2.1 seconds*. Princeton, NJ: Princeton University Press.

Hunt, H., J. Jinn, L. Jacobs, and R. Full. 2021. Acrobatic squirrels learn to leap and land on tree branches without falling. *Science* 373:697–700.

Hutto, D., M. Kirchhoff, and E. Myin. 2014. Extensive enactivism: Why keep it all in? *Frontiers in Neuroscience* 8:706.

Irwin, G. G., K. R. Williams, D. G. Kerwin, H. von Lieresund Wilkau, and K. M. Newell. 2021. Learning the high bar longswing: II. Energetics and the emergence of the coordination pattern, *Journal of Sports Sciences* 39(23):2698–2705. https://doi.org/10.1080/02640414.2021.1953829.

Jablonka, E., and M. J. Lamb. 2002. The changing concept of epigenetics. *Annals of the New York Academy of Sciences* 981(1):82–96.

Jacobs, J. 1993. *The death and life of great American cities*. New York: Random House.

Jantsch, E. 1980. *The Self-organizing universe*. Oxford: Pergamon.

Juarrero, A. 1991. Fail-safe versus safe-fail: Suggestions toward an evolutionary model of justice. *Texas Law Review* 69:1745–1777.

Juarrero, A. 1992. The message whose message it is that there is no message (in special Comparative Literature issue on *Foucault's Pendulum*, with commentary by U. Eco). *Modern Language Notes (MLN)* 107:892–894.

Juarrero, A. 1993. Des racines modernes aux rhizomes post-modernes (From modern roots to postmodern rhizomes). *Diogene* 163:29–48.

Juarrero, A. 1999. *Dynamics in action: Intentional behavior as a complex system.* Cambridge, MA: MIT Press.

Juarrero, A. 2004. Self-organization, individuation and identity. *Revue Internationale de Philosophie* 59:151–172.

Juarrero, A. and C. Rubino, eds. 2010. *Emergence, complexity and self-organization: Precursors and prototypes.* Boston: ISCE Publishing.

Juarrero-Roque, A. 1985. Self-organization: Kant's concept of teleology and modern chemistry. *The Review of Metaphysics* 39:107–135.

Jumper, J., R. Evans, A. Pritzel, et al. 2021. Highly accurate protein structure prediction with AlphaFold. *Nature* 596:583–589. https://doi.org/10.1038/s41586-021-03819-2.

Kauffman, S. 2014. Prolegomenon to patterns in evolution. *Biosystems* 123:3–8. https://doi.org/ 10.1016/j.biosystems.2014.03.004.

Kaufman, M. T., M. M. Churchland, S. Ryu, and K. V. Shenoy. 2014. Cortical activity in the null space: permitting preparation without movement. *Nature Neuroscience* 17(3):440–448. https://doi.org/10:1038/nn.3643.

Kelso, J. A. S. 1995. *Dynamic patterns: The self-organization of brain and behavior*. Cambridge, MA: MIT Press.

Kelso, J. A. S. 2008. Synergies: Atoms of brain and behavior. In *A multidisciplinary approach to motor control*, edited by D. Sternad. Heidelberg: Springer.

Kelso, J. A. S. 2009. Coordination dynamics. In *Encyclopedia of complexity and systems sciences*, edited by R.A. Meyers. Berlin: Springer-Verlag.

Kim, J. 1989. The myth of nonreductive materialism. *Proceedings and Addresses of the American Philosophical Association* 63(3):31–47.

Kim, J. 1998. *Mind in a physical world.* Cambridge, MA: MIT Press.

Kirschner, M. W., and J. C. Gerhart. 2005. *The plausibility of life: Resolving Darwin's dilemma.* New Haven, CT: Yale University Press.

Koch, C. 2004. *The quest for consciousness: A neurobiological approach.* Englewood, CO: Roberts and Company Publishers.

Koestler, A. 1968. *The ghost in the machine*. New York: Macmillan.

Kruesi, Liz. 2013. The cosmos' hidden scaffolding. *Astronomy.* https://astronomy.com/magazine/2013/07/the-cosmos-hidden-scaffolding.

Kurth, C. 2021. Disgust can be morally valuable. *Scientific American.* https://www.scientificamerican.com/article/disgust-can-be-morally-valuable/.

Kyselo, M. 2014. The body social: An enactive approach to the self. *Frontiers in Psychology* 5:986.

Kyselo, M., and W. Tschacher. 2014. An enactive and dynamical systems theory account of dyadic relationships. *Frontiers in Psychology* 5:452.

Ladyman, J., and D. Ross, with D. Spurrett and J. Collier. 2010. *Every thing must go*. Oxford: Oxford University Press.

Lange, M. 2007. Laws and meta-laws of nature: Conservation laws and symmetries. *Studies in History and Philosophy of Modern Physics* 38:457–481.

Lange, M. 2017. *Because without cause: Non-causal explanations in science and mathematics*. New York: Oxford University Press.

Laughlin, R., and D. Pines. 2000. The theory of everything. *Proceedings of the National Academy of Sciences* 97:28–31.

Lenton, T., and B. Latour. 2018. Gaia 2.0. *Science* 361(6407):1066–1068. https://doi.org/10.1126/science.aau0427.

Lenton, T. M., S. Daines, J. Dyke, A. Nicholson, D. Wilkinson, and H. Williams. 2018. Selection for Gaia across multiple scales. *Trends in Ecology and Evolution* 33(8):633–645. https://doi.org/10.1016/j.tree.2018.05.006.

Levins, R. 1973. The limits of complexity. In *Hierarchy theory*, edited by H. H. Pattee. New York: Braziller.

Lipscomb, B. J. B. 2021. *The women are up to something: How Elizabeth Anscombe, Philippa Foot, Mary Midgley, and Iris Murdoch revolutionized ethics*. Oxford: Oxford University Press.

Lustgarten, A. 2022. The Barbados rebellion. *New York Times Magazine*. https://www.nytimes.com/interactive/2022/07/27/magazine/barbados-climate-debt-mia-mottley.html.

MacIntyre, A. 2007. *After virtue*. Notre Dame, IN: University of Notre Dame Press.

Mante, V., D. Sussillo, K. V. Shenoy, and W. T. Newsome. 2013. Context-dependent computation by recurrent dynamics in prefrontal cortex. *Nature* 503(7474):78–84. https://doi.org/10.1038/nature12742.

Marshall, M. 2021. How water makes life possible. *NewScientist*, January 23, 15.

Martone, Robert. 2020. Music synchronizes the brains of performers and their audience. *Scientific American*. https://www.scientificamerican.com/article/music-synchronizes-the-brains-of-performers-and-their-audience/?print=true.

Mason, P. H., J. F. Dominguez, B. Winter, and A. Grignolio. 2014. Hidden in plain view: Degeneracy in complex systems. *Biosystems* 128:1–8. https://doi.org/10.1016/j.biosystems.2014.12.003.

Maturana, H., and F. Varela. 1979. *Autopoiesis and cognition: The realization of the living*. Dordrecht, Netherlands: D. Reidel Publishing.

Maury, C. P. J. 2018. Amyloid and the origin of life: self-replicating catalytic amyloids as prebiotic informational and protometabolic entities. *Cellular and Molecular Life Sciences* 75:1499–1507. https:/doi.org/10.1007/s00018-18-2797.9.

Maynard Smith, J., and E. Szathmary. 1995. *The major transitions in evolution*. Oxford: Oxford University Press.

Mazzucato, M. 2020. *The value of everything*. New York: Public Affairs Books.

McCulloch, W. 1945. A heterarchy of values determined by the topology of nervous nets. *Bulletin of Mathematical Biophysics* 7(2):89–93.

McMullin, B. 1999. Some remarks on autocatalysis and autopoiesis. http://www.eeng.dcu.ie/-mcmullin/.

McMullin, B. 2000. Remarks on autocatalysis and autopoiesis. *Annals of the New York Academy of Sciences* 901:163–174.

Medzhitov, R. 2021. The spectrum of inflammatory responses. *Science* 374(6571):1070–1075. https://doi.org/10.1126/science.abi5200.

Metzinger, T., ed. 2000. *Neural correlates of consciousness: Empirical and conceptual questions*. Cambridge, MA: MIT Press.

Mitchell, M. 2021. Why AI is harder than we think. arXiv:2104.12871v2 [cs.AI].

Montevil, M., and M. Mossio. 2015. Biological organization as closure of constraints. *Journal of Theoretical Biology* 372:179–191.

Moreno A., and M. Mossio. 2015. *Biological autonomy: A philosophical and theoretical inquiry. New York: Springer.*

Mossio, M. 2013. Closure, causal. In *Encyclopedia of systems biology*, edited by W. Dubitzky, O. Wolkenhauer, K.-H. Cho, and H. Yokota, 415–418. New York: Springer.

Murphy, N., and W. Brown. 2007. *Did my neurons make me do it*. Oxford: Oxford University Press.

Murphy, N., G. F. R. Ellis, and T. O'Connor, eds. 2009. *Downward causation and the neurobiology of free will*. New York: Springer.

Nagel, T. 1986. *The view from nowhere*. Oxford: Oxford University Press.

Nicolis, G., and I. Prigogine. 1977. *Self-organization in nonequilibrium systems*. New York: Wiley.

Noe, A. 2010. *Out of our heads: Why you are not your brain, and other lessons from the biology of consciousness*. New York: Hill and Wang.

Nordholm, S., and G. B. Bacskay. 2020. The basics of covalent bonding in terms of energy and dynamics. *Molecules* 25:2667. https://doi.org/10.3390/molecules25112667.

Pascal, R., and A. Pross. 2015. Stability and its manifestation in the chemical and biological words. *ChemComm* 51:16160–16165.

Pattee, H. H. 1972a. Laws and constraints, symbols and languages. In *Towards a theoretical biology*, edited by C. H. Waddington. Vol. 4. Edinburgh: Edinburgh University Press.

Pattee, H. H. 1972b. The evolution of self-simplifying systems. In *The relevance of general systems theory*, edited by E. Laszlo, 31–42. New York: Braziller.

Pattee, H. H. 1973. The physical basis and origin of hierarchical control. In *Hierarchy theory: The challenge of complex systems*. New York: Braziller.

Pattee, H. H. 1978. Biological systems theory: Descriptive and constructive complementarity. In *Applied general systems research*, edited by G. J. Klir, 511–520. New York: Plenum.

Pattee, H. H. 1982. Cell psychology: An evolutionary approach to the symbol-matter problem. *Cognition and Brain Theory* 4:325–341.

Patten, B., and G. Auble. 1980. Systems approach to the concept of niche. *Synthese* 43(1):155–181.

Pearl, J. 2009. *Causality: Models, reasoning and inference*. Cambridge: Cambridge University Press.

Pendleton-Jullian, A. 2020. Coming of Age: From frameworks and theories of change to scaffolds for ecologies of change. In *Cynefin: Weaving sense-making into the fabric of our world,* edited by R. Greenberg and B. Bertsch. Singapore: Cognitive Edge.

Pendleton-Jullian, A., and J. S. Brown. 2016. *Pragmatic imagination*. San Francisco: Blurb Press.

Perry, J. 2002. *Identity, personal identity, and the self*. Indianapolis, IN: Hackett.

Place, U. T. 1956. Is consciousness a brain process? *British Journal of Psychology* 47:44–50.

Plaut, D. C., and T. Shallice. 1993. Deep dyslexia: A case study of connectionist neuropsychology. *Cognitive Neuropsychology* 10:377–500.

Polger, T. W., and L. A. Shapiro. 2016. *The multiple realization book*. New York: Oxford University Press.

Prigogine, I., and I. Stengers. 1984. *Order out of chaos: Man's new dialogue with nature*. Toronto: Bantam.

Pross, A. 2012. *What is life?: How chemistry becomes biology*. New York: Oxford University Press.

Pross, A., and R. Pascal. 2013. The origin of life: What we know, what we can know and what we will never know. *Open Biology* 3. https://doi.org/10.1098/rsob.120190.

Putnam, H. 1967. Psychological predicates. In *Art, mind, and religion*, edited by W. H. Capitan and D. D. Merrill, 37–48. Pittsburgh: University of Pittsburgh Press.

Putnam, H. 1975. The meaning of "meaning." *Language, Mind, and Knowledge* 7:131–193.

Rampino, M. R., K. Caldeira, and Y. Zhu. 2021. The earth has a pulse: A 27.5 Myr underlying cycle in coordinated geological events over the last 260 Myr. *Geoscience Frontiers 12(6)*. https://doi.org/10.1016/j.gsf.2021.101245.

Ravilious, K. 2021. The last human. *NewScientist* 27:41.

Rayleigh, Lord. 1916. On the convective currents in a horizontal layer of fluid when the higher temperature is on the underside. *Philosophical Magazine* 32(192):529–546.

Regalado, A. 2019. Chinese scientists have put human brain genes in monkeys—and yes, they may be smarter. MIT Technology Review. https://www.technologyreview.com/2019/04/10/136131/chinese-scientists-have-put-human-brain-genes-in-monkeysand-yes-they-may-be-smarter/.

Ricci-Tam, C., I. Ben-Zion, J. Wang, J. Palme, and A. Li. 2021. Decoupling transcription factor expression and activity enables dimmer switch gene regulation. *Science* 372(6539):292–295. https://doi.org/10.1126/science.aba/7582.

Roque, A. J. 1987. Does action theory rest on a mistake? *Philosophy Research Archives* 13:587–612.

Rorty, R. 1982. *Consequences of pragmatism: Essays 1972–1980*. Minneapolis: University of Minnesota Press.

Rosen, R. 1985. *Anticipatory systems*. Pergamon Press.

Rosenberg, Alexander. 2001. On multiple realization and the special sciences. *The Journal of Philosophy* 98:365–373.

Rosenblueth, A., N. Wiener, and J. Bigelow. 1943. Behavior, purpose and teleology. *Philosophy of Science* 10:18–24.

Ross, R. 1916. An application of the theory of probabilities to the study of *a priori* pathometry, Part 1. *Proceedings of the Royal Society Series A* 92(638): 204–230.

Rovelli, C. 2018. *The order of time*. New York: Riverhead.

Rowlands, M. 2010. The mind embedded. In *The new science of the mind: From extended mind to embodied phenomenology*. Cambridge, MA: MIT Press.

Ruiz-Mirazo, K., and A. Moreno. 2004. Basic autonomy as a fundamental step in the synthesis of life. *Artificial Life* 10(3):235–259.

Ryle, G. 1949. *The concept of mind*. Chicago: University of Chicago Press.

Saberi, M., H. Hamedmoghadam, M. Ashfaq, et al. 2020. A simple contagion process describes spreading of traffic jams in urban networks. *Nature Communications* 11:1616. https://doi.org/10.1038/s41467-020-15353-2.

Salthe, S. 1991. Varieties of emergence. *World Futures* 21(2):69–73.

Salthe, S. 1980. Robustness, reliability and multiple determinism in science: The nature and variety of a powerful family of problem-solving heuristics. In *Knowing and validating in the social sciences: A tribute to Donald T. Campbell*, edited by M. Brewer and B. Collins. San Francisco: Jossey-Bass.

Salthe, S. 1985. *Evolving hierarchical systems*. New York: Columbia University Press.

Salthe, S. 1993. *Development and evolution: Complexity and change in biology*. Cambridge, MA: MIT Press.

Salthe, S. 2001. Summary of the principles of hierarchy theory. https://www.nbi.dk/~natphil/salthe/Summary_of_the_Principles_o.pdf.

Salthe, S. 2010. Maximum power and maximum entropy production: Finalities in nature. *Cosmos and History: The Journal of Natural and Social Philosophy* 6(1):114–121.

Salthe, S. 2012. Hierarchical structures. *Axiomathes* 22:355–383. https://doi.org/10.1007/s10516-012-9185-0.

Salthe, S. N. 2015. What actually is a living system materially? *Biological Theory 11:50–55*. https://doi.org/10.1007/s13752-015-0230-2.

Salthe, S. 2018. Perspectives on natural philosophy. *Philosophies* 3(23):35–44. https://doi.org/10:3390/philosophies3030023.

Sancar, A., and R. N. Van Gelder. 2021. Clocks, cancer and chronochemotherapy. *Science* 371(6524):42. https://doi.org/10.1126/science.abb0738.

Sawyer, R. 2005. *Social emergence: Societies as complex systems*. Cambridge: Cambridge University Press.

Schelling, T. C. 1971. Dynamic models of segregation. *Journal of Mathematical Sociology* 1(2):143–186.

Searle, J. 1980. Minds, brains and programs. *Behavioral and Brain Sciences* 3:417–457.

Shannon, C., and W. Weaver. 1949. *The mathematical theory of communication.* Champaign: University of Illinois Press.

Shapere, D. 1982. The concept of observation in science and philosophy. *Philosophy of Science* 49(4):485–525.

Shapiro, L. A. 2000. Multiple realizations. *The Journal of Philosophy* 97(12):635–654. https://doi.org/10.2307/2678460.

Shaw, R. and M. T. Turvey. 1981. Coalitions as models for ecosystems: A realist perspective on perceptual organization." In *Perceptual organization,* edited by M. Kubovy and J. Pomerantz, 343–416. Hillsdale: Lawrence Erlbaum.

Sheehy, J. 2015. There is no now. *Communications of the ACM* 58:36–41.

Shenoy, K. V., M. T. Kaufman, M. Sahani, and M. M. Churchland. 2011. A dynamical systems view of motor preparation: Implications for neural prosthetic system design. *Progress in Brain Research* 192:33–58.

Shiller, R. J. 2019. *Narrative economics: How stories go viral and drive major economic events*. Princeton, NJ: Princeton University Press.

Simon, H. 1969. The architecture of complexity. In *The sciences of the artificial.* Cambridge, MA: MIT Press.

Simon, H. 1973. The organization of complex systems. In *Hierarchy theory: The challenge of complex systems*, edited by H. H. Pattee. New York: Braziller.

Slaby, J., and G. Gallagher. 2015. Critical neuroscience and socially extended minds. *Theory, Culture and Society* 32(1):35–59.

Smart, J. J. C. 1959. Sensations and brain processes. *Philosophical Review* 68:141–156.

Snowden, D. 2015. Naturalizing sensemaking. In *Informed by knowledge: Expert performance in complex situation*, edited by K. Mosier and U. Fischer, 223–234. New York: Routledge.

Snowden, D. 2020. Cynefin: A tale that grew in the telling. In *Cynefin: Weaving sense-making into the fabric of our world*, edited by R. Greenberg and B. Bertsch. Singapore: Cognitive Edge_The Cynefin Co.

Snowden, D., and A. Pendleton-Jullian. 2020. Scaffolding. https://cdn.cognitive-edge.com/wp-content/uploads/sites/7/2020/04/22082941/c-CognitiveEdge_Scaffolding.pdf.

Steiner, P. 2019. Brain fuel utilization in the developing brain. *Annals of Nutrition and Metabolism* 19(75, suppl. 1):8–18.

Stenz, L., D. S. Schechter, S. R. Serpa, and A. Paoloni-Giacolino. 2018. Intergenerational transmission of DNA methylation signatures associated with early life stress. *Current Genomics* 19(8):665–675.

Subbaraman, Ni. 2021. First monkey-human embryos reignite debate over hybrid animals. *Nature* 592:497.

Thelen, E., and L. B. Smith. 1994. *A dynamic systems approach to the development of cognition and action*. Cambridge, MA: MIT Press.

Thom, R. 1989. *Structural stability and morphogenesis: An outline of a general theory of models*. Reading, MA: Addison-Wesley.

Thomas, A., B. Woo, D. Nettle, E. Spelke, and R. Saxe. 2022. Early concepts of intimacy. *Science* 375(6578):311–315.

Tononi, G. 2012. *Phi: A voyage from the brain to the soul*. New York: Pantheon.

Tononi, G. 2008. Consciousness as integrated information: A provisional manifesto. *Biological Bulletin* 215:216–242.

Toulmin, S. 1990. *Cosmopolis: The hidden agenda of modernity*. New York: Free Press.

Turbek, S., M. Browne, A. S. Di Giacomo. 2021. Rapid speciation via the evolution of pre-mating isolation in the Iberá Seedeater. *Science* 371:6536.

Turing, A. 1936. On computable numbers, with an application to the Entscheidungsproblem. *Proceedings of the London Mathematical Society* (Series 2) 42:230–265.

Turner, T. L., M. W. Hahn, and S. V. Nuzhdin. 2005. Genomic islands of speciation in *Anopheles gambiae*. *PLoS Biology 3*:e285.

Turvey, M. 1990. Coordination. *American Psychologist* 45:938–953.

Turvey, M. T., K. Shockley, and C. Carello. 1999. Affordance, proper function, and the physical basis of perceived heaviness. *Cognition* 17:B17–B26.

Ulanowicz, R. 1997. *Ecology: The ascendent perspective*. New York: Columbia University Press.

van Gulick, R. 1993. Who's in charge here and who's doing all the work? In *Essays on mental causation*, edited by J. Heil and A. Mele, 233–256. Oxford: Oxford University Press.

van Gulick, R. 2004. Higher-order global states (HOGS): An alternative higher-order model of consciousness. In *Higher-order theories of consciousness: An anthology*, edited by R. J. Gennaro. Amsterdam: John Benjamins.

Van Orden, G. C., J. G. Holden, and M. T. Turvey. 2003. Self-organization of cognitive performance. *Journal of Experimental Psychology: General* 132(3):331.

Varela, F., and E. Thomson. 2003. Neural synchronicity and the unity of mind: A neurophenomenological perspective. In *The unity of consciousness: Binding, integration, and dissociation*, edited by A. Cleermans. Oxford: Oxford University Press.

Voosholz, J., and M. Gabriel, eds. 2021. *Top-down causation and emergence*. Cham, Switzerland: Springer.

Vygotsky, L. 1978. *Mind in society: The development of higher psychological processes*. Cambridge, MA: Harvard University Press.

Waddington, C. 1968–1972. *Towards a theoretical biology*. 4 vols. Edinburgh: Edinburgh University Press.

Wagner, A. 2005. *Robustness and evolvability in living systems*. Princeton, NJ: Princeton University Press.

Wagner, A. 2014. *Arrival of the fittest: Solving evolution's greatest puzzle*. New York: Penguin.

Weiss, P. 1971. The basic concept of hierarchic systems. In *Hierarchically organized systems in theory and practice*, edited by P. Weiss. New York: Hafner.

Wheeler, J. A. 2000. *Geons, black holes and quantum foam: A life in physics*. Rev. ed. New York: W. W. Norton.

Wilson, C. 2020. The mystery of mistletoe's missing genes. *QuantaMagazine*. https://www.quantamagazine.org/the-mystery-of-misteltoes-missing-genes-20201221/.

Wimsatt, W. C. 1974. Reductive explanation: a functional account. *Proceedings of the Biennial Meeting of the Philosophy of Science Association* 1974:671–710. http://www.jstor.org/stable/495833.

Wimsatt, W. C. 1976. Reductionism, levels of organization, and the mind-body problem. In *Consciousness and the brain: A scientific and philosophical inquiry*, edited by G. G. Globus, G. Maxwell, and I. Sarodnik. New York: Plenum.

Wimsatt, W. C. 2001. Generative entrenchment and the developmental systems approach to evolutionary process. In *Cycles of contingency: Developmental systems and evolution*, edited by S. Oyama, R. Gray, and P. Griffiths. Cambridge, MA: MIT Press.

Wimsatt, W. C. 2007. *Reengineering philosophy for limited beings*. Cambridge, MA: Harvard University Press.

Woods, D. D. 2016. Resilience as graceful extensibility. In *IRGC resource guide on resilience*. Lausanne: EPFL International Risk Governance Center. v29-07-2016. https://www.irgc.org/riskgovernance/resilience/.

Woods, D. D. 2018. The theory of graceful extensibility. *Environment Systems and Decisions* 38:433–457.

Wright, L. 1976. *Teleological explanations*. Berkeley: University of California Press.

Yang, T., M. Hudson, and N. Afshordi. 2020. How dark are filaments in the cosmic web. https://arxiv.org/abs/2001.10943.

Yu, B. M., A. Afshar, G. Santhanam, S. I. Ryu, K. V. Shenoy, and M. Sahani. 2006. Extracting dynamical structure embedded in neural activity. *Advances in Neural Information Processing Systems* 18:1545–1552.

Zhou, L., G. B. Melton, S. Parsons, and G. Hripcsak. 2005. A temporal constraint structure for extracting temporal information from clinical narrative. *Journal of Biomedical Informatics* 39:424–439.

Index

Page numbers in italic indicate figures.